ORDINARY IN BRIGHTON?:
LGBT, ACTIVISMS AND THE CITY

For all those who shared their stories, time and passion investing in the research and social change it sought to achieve.

Ordinary in Brighton?: LGBT, Activisms and the City

KATH BROWNE
University of Brighton, UK

LEELA BAKSHI
Research Activist, UK

Routledge
Taylor & Francis Group

LONDON AND NEW YORK

First published 2013 by Ashgate Publishing

2 Park Square, Milton Park, Abingdon, Oxon OX14 4RN

711 Third Avenue, New York, NY 10017, USA

Routledge is an imprint of the Taylor & Francis Group, an informa business

First issued in paperback 2016

British Library Cataloguing in Publication Data
A catalogue record for this book is available from the British Library.

The Library of Congress has cataloged the printed edition as follows:
Browne, Kath.
  Ordinary in Brighton? : LGBT, activisms and the city / by Kath Browne and
Leela Bakshi.
     pages cm
  Includes bibliographical references and index.
  ISBN 978-1-4724-1294-2 (hardback)
1. Sexual minorities--England--Brighton--Social
conditions. 2. Sexual minorities--Government policy--England--Brighton. I. Bakshi,
Leela. II. Title.

  HQ73.3.G72B753 2013
  306.7609422'56--dc23

                                                                2013005638

ISBN 978-1-4724-1294-2 (hbk)
ISBN 978-1-138-25122-9 (pbk)

Contents

List of Figures and Tables

Figures

Tables

Preface

This book reflects on learning from the stories, hopes and views entrusted to the Count Me In Too (CMIT) research project. This project sought to work towards improving Lesbian, Gay, Bisexual and Trans (LGBT) lives through working across community, university and policy making sectors. In addition to producing a range of outputs orientated to practitioners and policy makers and community summaries providing briefer information on key findings (all available at www. countmeintoo.co.uk), CMIT findings were disseminated through publication in academic journal articles and conference presentations listed on the project website (see also Appendix 1). This book further develops this thinking, linking and extending the research literature whilst also honouring the contributions of those who took part.

The topics explored in these outputs reflect that the project fed into and from activisms and agendas that were prominent in LGBT Brighton at the time of the research. This included themes explored in the chapters of this book (the scene, bi lives and trans lives, safety, pride) as well as those identified for further themed analysis by service providers and community groups during the project (including mental health, domestic violence and drugs and alcohol). As these became foregrounded, other agendas were de-centred, which in other contexts we and other scholars might see as requiring attention. For example, questions of race and ethnicity were given considerable attention in the research design and process and these identities were highlighted as pertinent by the steering group. The collective undertaking of the project sought the participation of those who occupy these and other positions (including asylum seekers and people of faith, as well as a range of other identities) (see Browne et al. 2009 for a critical examination of this). Focus groups were undertaken specifically targeting participation of Black and Minority Ethnic (BME) people and all quantitative analysis explored ethnicity as a variable (see Appendix 1 for details of the research). However, key findings in this research, identified in part by LGBT people through participatory processes, did not deal in depth with race and ethnicity. This is not to say that whiteness, racism and the intersectionalities of these with gender and sexual identities (see Appendix 2) are not important or were not relevant in understanding Brighton. Indeed, we consider race at different points in this text. Rather, because of the participatory way in which the CMIT research operated, the research team prioritised issues that were felt at that time to have the most potential to productively work towards the aim of progressing social change for LGBT people in the city. We recognise that particular social differences may be highlighted when engaging with data that emerges from a participatory process, to the neglect of those that are of crucial importance in other contexts. In this case academic agendas that demand analysis of the intersections of race and ethnicity are important, but these were not the focus of those working pragmatically at the front line of LGBT equalities. However, with this absence of data, we cannot centralise this as a point of analysis and anticipate that this will be seen as a failing of the research and an absence in this text. This study, as in all research, has silences and absences; here we name one that we not only recognise as a potential point of critique, but also as an issue that we both experience personally; one of us identifies as BME and the other as a first generation Irish immigrant. Whilst we are uncomfortable with this absence, we are unapologetic that this results from not 'pushing' our agendas in order to produce material that we 'need' for our academic endeavours and this book. We hope that other narrations of LGBT organising and activism that attend to identities neglected in this work will be enlivened by this book.

Ordinary in Brighton? seeks to challenge the academic audience for whom it is framed, to consider the process of formation of knowledge. We use extended quotes drawn from the Count Me In Too participatory research process, to contest the authorial voice of the scholar and offer insights from those who participated and formed the research. We at times treat these as 'fact', reflecting the expertise of those we spoke to. While some might read this as descriptive, we argue that displacing the scholar has profound and necessary implications for the creation of knowledge. This is also achieved through a co-authorship between an activist researcher (Leela) and an academic (Kath).

In displacing and re-placing academic conventions and challenging the god-trick of academic authorship through the rich tradition of feminist discussions of positionalities, it is important to acknowledge who we 'are' as authors, researchers and activists and why we took part in this project (see, for example, Haraway 1991, Rose 1995). Leela says:

> I really wanted to write. Writing the book also presented a logical next step for (my) activism through the Count Me In Too project, where there was more learning to explore and a need for a vehicle to disseminate ideas arising in the project. It is important to me how writing undresses me from less privileged statuses – woman, black, lesbian, feminist, mixed race, bit wobbly with mental health and other stuff. I don't have to deal with the interactions around these identities when I'm communicating through writing.

For Kath:

> writing this book was both a necessity (outputs from the research), and had massive personal investment. Both living and enjoying this city and not feeling 'marginalised' by my difference (being Irish and a recent immigrant), gave me the opportunity and drove me to understand what 'gay Brighton' could do to those who did not have the privileges I enjoy. Throughout this process, and now writing with Leela, my values and judgements have been questioned in

ways that can be described as inspirational and enriching, as well as difficult, challenging and at times frustrating!

We believe that our position as women in same-sex relationships, who define themselves as lesbians, enabled both of us, but Kath in particular, to gain access in ways that would not have been possible for a straight or male researcher. At the time of the research we were each engaged with local LGBT activism: Leela's work as a trustee of Spectrum (2005–2010) led to her involvement with Count Me In Too and Kath's position as a trustee of Pride (2005–2010) impacted on the research and the writing of this book, including chapters and sections on Spectrum (see Chapter 2 and 6) and Pride (see Chapter 8). However, we sought to explore the nuances, contradictions and tensions that emerged in the data regarding both of these organisations in self-critical and reflexive ways. We have also used our friendship to critically examine our emotional attachments and investments to various aspects of gay Brighton. Similarly our findings around lesbians and bi women in the second part of the book were unexpected but closely felt.

Often community and academic outputs are separated, presuming diverse audiences with differing interests (theory versus practice) and abilities (accessibility versus complexity and nuance). Whilst we set out with the intention of writing across academic/activist audiences, our initial drafts were seen as failing to realise this and not catering appropriately for either. We moved to privilege academic thought and content in order to make the book 'publishable' within an academic context, thus developing support for 'soundness' of the research upon which it is based. Neither of us were entirely comfortable with this shift but saw it as expedient, fulfilling the aims of reaching academic and activists with interest in research and theorisation. This highlights some of the conventions, boundaries and knowledge creation techniques that continue to privilege particular 'experts'. We hope however that this book goes some way to challenging conventions of academic knowledge creation and opens up space for further experiments and innovations that rework both academic standards and marketability.

Writing together is another way to challenge imbalances of power/privilege and writing in our roles as academic/activist researcher requires the negotiation of a mosaic of special skills and competencies. Although it did not feel for either of us like negotiating imbalances of power/privilege, we agree that the love for what we're doing eclipses (and obscures) the complexity of who is supporting who with what. Specifically, we sat down and discussed ideas and what might go in the book, with each of us given particular tasks to write on. Kath drafted chapters and Leela wrote 'patchwork pieces' which were often sewn together with other pieces Kath located and prepared for the project. A structure was then applied that Kath conceived and then executed. We took turns to review, edit and rewrite sections of chapters until we had ironed our work into the finished book. Through the process, chapters were sent to readers, academic and activists whose comments enabled us to rework and develop our ideas and writing. Readers and reviewers gave us invaluable feedback regarding the directions of the book and focusing our arguments. This altered the shape and content of the book. Nonetheless the book reflects our interests, theoretical impulses and orientations. We see this writing as one possible reading of the material, context and conceptualisations, and hope that it serves to open up, rather than foreclose, alternative engagements with LGBT lives and activisms.

Kath Browne and Leela Bakshi
November 2012

Acknowledgements

We would like to begin by thanking the many individuals, organisations and services that helped make Count Me In Too possible. Thank you to everyone who completed a questionnaire or attended a focus group for your time and trust. Thank you to all who participated in a consultation or feedback event,, or an individual interview, particularly: Yvonne Barker, Jo Barringer, Robert James Clothier, Dana Cohen, Mary Evans, Sheila McWattie, Sam Thomas, Persia West, Natalie Woods and 10 others. We hope that all of your stories will make a lasting difference.

This project, as a community–university partnership, was possible because of the amazing work undertaken by Spectrum and particular thanks are owed to Arthur Law, coordinator of Spectrum LGBT community forum for the duration of the project who worked tirelessly for positive social change for LGBT people. Thanks to Jason Lim, Nick McGlynn and Sharif Mowlabocus for their extensive and fabulous work on the Count Me In Too project.

Thank you to everyone who volunteered with the project at each stage, including the Count Me In Too Community Steering Group, Action Group, Monitoring Group and the Count Me In Too Analysis Groups. The project couldn't have happened without you. Thank you to the local organisations that provided support in-kind by making available staff and resources to work with the project. Your collaboration has enhanced the project's work immeasurably.

Thank you to the following organisations and services who provided financial support throughout the project: the University of Brighton, Brighton and Sussex Community Knowledge Exchange, NHS Brighton and Hove (formerly Brighton and Hove City Teaching Primary Care Trust), Brighton and Hove City Council, CUPP, South East Coastal Communities (HEFCE), Brighton and Hove City Council Social Care and Housing, the Partnership Community Safety Team, Sussex Police, the Sussex Partnership NHS Foundation Trust, the Drug and Alcohol Action Team.

We want to thank everyone else who in various ways has helped to make this research happen: including all who designed, debated and contributed questions to the questionnaire, all who offered comments and help on the process, all who attended stakeholder and community meetings and feedback events.

Thank you to all of those who read and offered comments on earlier versions of the chapters in this book, particularly Michelle Buckley, Andrew Church, Camel Gupta, Phil Hubbard, Sharif Mowlabocus, Catherine Nash, David Nash, Julie Nichols, Cassandra Smith, Georgina Voss and the manuscript and proposal reviewers. The comments we received on earlier versions of this book vastly improved the focus and direction of the book and we are deeply indebted to those who gave their time and advice so generously. Thank you to those who gave us feedback at various conferences and presentations, including the European Geographies of Sexualities Conference (2011); Royal Geographical Society/ Institute of British Geographers Annual Conference (2011, 2012); Association of American Geographers Annual Conference (2011) and the staff and students at the University of British Columbia, Vancouver. Thank you to Craig Haslop and Cassandra Smith for all of their help in getting this book to publication. Thanks to Gscene for the kind permission to reprint the 'Map of the Gay Scene'.

Leela would like to thank Val for her love and support through our journey and with the extra baggage I throw on board – like 'next, I'm going to co-write a book …'. Thank you to the LGBT activists that I have met and worked with. You are each an inspiration and role model to me. Thank you to my friends and family for believing in me, making it possible for me to move forwards. Thank you to Mr Brown, my geography teacher and form tutor at secondary school, who taught me about loving geography and living as myself, which (eventually) led me to co-write this book.

Kath would like to thank Donna Imrie-Browne and all of those who have supported and listened to her throughout the Count Me In Too process. I am grateful for all who acted as mentors and patiently taught me the complexities and nuances of working in and with LGBT Brighton. I feel deeply honoured to have been part of this research and to have met some of the most amazing people who shared their lives with us. I am deeply indebted to LGBT activists who passionately work towards making life better.

Thanks to Andrew Weekes for permission to use the cover image.

Chapter 1
Equalities, Cities and Ordinariness

Introduction

The early twenty-first century saw far-reaching and unprecedented legislative changes in places such as the UK, Canada, Scandinavia and other countries within the European Union that made some lesbian, gay, bisexual and trans (LGBT)[1] people 'ordinary', rather than sexual/gender deviants (see Stychin 2003, 2005, Richardson 2004, 2005, Richardson and Monro 2012). The first decade of the new millennium in the UK culminated in the 2010 Equality Act, described by Cooper (2011: 4) as 'the high water mark of anti-discrimination reform in Britain to date'.

The profound shift in focus in terms of sexual (and to an extent gender) citizenship in England mirrored changes in political power in the country, which moved from a centre-right Conservative government from 1979–1997 to the left of centre politics of New Labour from 1997–2010. Under the Conservatives, sexual and gender dissidence was read as beyond the boundaries of the nation state, and lesbians and gay men remained the 'enemy within' (Carabine and Monro 2004: 28, Bell and Binnie 2000). Lesbian and gay sexualities work in certain local government authorities in the 1980s met with a Conservative backlash, leading to the 1992 section 28 of the Local Government Act, banning the 'promotion of homosexuality' (Cooper 1994, Monro 2006, 2007, 2010). The tide began to change in the mid to late 1990s, when there was a legal and cultural shift away from considering lesbians and gay lives as (im)moral, towards conceptualisations of these as social identities with associated 'problems of bias, prejudice and discrimination' (Cooper 2006: 936).

Legal change was accompanied by the formation of national equality bodies such as the Government Equalities Office and the Equality and Human Rights Commission, as well as other NGOs and local organisations (Richardson and Monro 2012, McDermott 2011). During the first decade of the twenty-first century, targeted programmes and initiatives for LGBT people came into operation and were 'state funded'. The push of these 'exterior' legislative changes has, in McNulty et al.'s (2010) research, resulted in drivers of social change where LGBT work is embedded into statutory contexts. For LGBT people (along with a range of other groupings) there was an expanded consideration of 'social exclusion' moving from the sole remit of poverty and deprivation to inclusion in a broader sense

---

1   There has been an extensive discussion of the use of this term, it's constructedness as well as exclusions. These debates have been well addressed elsewhere. In Appendix 1 we detail some of these discussions and justify our deployment of the term LGBT.

Table 1.1 Timeline of key equalities gains

Date	Event
1997	New Labour (centre-left political party) elected
1998	Crime and Disorder Act requires community consultation
1999	Right to gender reassignment surgery paid for by the National Health Service achieved following a European Court ruling
1999	The Sex Discrimination (Gender Reassignment) Regulations 1999 extended the Sex Discrimination Act (1975) to make it unlawful to discriminate on grounds of gender reassignment, in the areas of employment and vocational training
2000	Local Government Act required local authorities to address social exclusion and made consultation with local communities a legal imperative in decision making processes
2000	Removal of the ban for lesbians and gay men in the armed forces
2001	Equalising the age of consent (16)
2003	Repeal of section 28 (banning the promotion of homosexuality in schools)
2003	Workplace protections legislated for sexual orientation
2004	Civil Partnership Act passed (first Civil Partnership 2005)
2004	Gender Recognition Act allows people to change their legal gender
2006	Equality Act 2006 includes the Gender Equality Duty, which places an obligation on public bodies to pay due regard to the need to address and eliminate the unlawful discrimination and harassment of transsexual people in employment, related fields and vocational training (including further and higher education) and in the provision of goods, facilities and services
2007	Equality Act makes it illegal to discriminate on the grounds of sexual orientation in the provision of goods and services
2008	New criminal offense, incitement to homophobic hatred, established
2008	Human Fertilisation and Embryology Act 2008, gives better legal recognition to same-sex parents
2008	Sex Discrimination (Amendment of Legislation) Regulations 2008 extended the Sex Discrimination Act to make it unlawful to discriminate on grounds of gender reassignment in the provision of goods, facilities and services
2010	Equality Act, which made sexual orientation and gender identity a protected characteristic with duties placed on public bodies to address discrimination and foster positive relations between diverse groups, including LGBT/non-LGBT

Sources: Stonewall 2013, Equalities Human Rights Commission 2013.

(Craig and Taylor 2002). This coincided with the Local Government Act (2000) which gave local state authorities a duty to prepare strategies to improve the social, environmental and economic wellbeing in their areas, 'identifying local actions to improve the quality of life for all sections of their community and with increasing consultation and participation' (Carabine and Monro 2004: 323). Thus, alongside UK equalities legislation addressing discrimination on the basis of gender and sexual difference, legislation in England moved to oblige local governments and statutory services[2] to proactively cater for their 'diverse' local populations. This included an imperative to consult with minority groups more robustly than before and, in cities such as Brighton, this included LGBT people. Here, this population was considered the largest 'minority' in the city (see also Moran et al. 2004). From the late 1990s, the parallel agendas of activists and statutory services presented opportunities for new modes of engagement. The supposedly common aspiration to achieve equality for LGBT people and deliver a social inclusion agenda gave rise to increased possibilities for activists, practitioners and policy makers to work together, often through partnership working[3] by state and community sectors. Service providers at times 'needed' activists or at least 'the community' in order to action their remit to consult with 'local communities' (see Chapters 6 and 7).

While some may see these changes as progressive, New Labour's 'reformers' have come under critical scrutiny in scholarly arenas. There are a range of different critiques, including pointing to the creation of new normalisations and liberalising equalities. Moreover, Moran et al. (2004) note that the effects of legislative change were 'concentrated and limited'. Therefore, work in this area was geographically uneven with continuing evidence of homophobic (as well as biphobic and transphobic, see Chapter 5) cultures, as well as 'pro-active LGB equalities work' (Monro 2010: 1001, Richardson and Monro 2012).

---

2   Local authorities in the UK are funded by taxes on the local population and by central government. They elect local members (councillors) every four years as well as employing paid workers. They are charged with managing infrastructure and the majority of welfare provision (including housing and income support) (Richardson and Monro 2012). Statutory services are services that are statutorily required but are not only located in local authorities, and include the police and health services. Statutory services are variously referred to in this book as public sector services and include mainstream services.

3   Partnership working is explored through work between voluntary and community groups and state providers. As we contend in Chapter 6, the state/non-state binary is unhelpful in understanding these partnerships, and the state is instead considered relationally. Partnership working in the UK context differs from that in other places, where private investors and philanthropy are key to the provision of services such as housing and health. In the UK, during the era in question, these were state provided and administered. Thus, changes to equalities legislation were used to make rights claims from those who were representing 'the state' and offered possibilities in affecting the provision for some of the most marginalised groups of LGBT people, particularly in Brighton where LGBT people were seen to be a 'significant minority' in the city.

Ordinary in Brighton? explores the changes in urban life for LGBT people during this period, the first decade of the twenty-first century. Examining Brighton as a city where LGBT populations supposedly became ordinary in the 1990s and 2000s, we critically question the extent to which urbanities can serve as landscapes of inclusion and equality, and examine the unevenly felt consequences of assimilation and inclusion in what is colloquially known as the UK's gay capital (www.gaybrighton.co.uk). The book offers the first large-scale examination of the impact of the UK equities legislation on lesbian, gay, bi and trans (LGBT) people themselves and on the nature of LGBT political activism within places such as Brighton, with its long history as a location favoured by LGBT people.

In this book we take a spatial perspective on questions of inclusion, showing how place creates sexual and gendered lives and activisms, through urban imaginings/representations that inform political possibilities. We will show how LGBT exclusions from the nation-state need to be augmented with explorations of the contingencies of localities because these play a part in creating social difference. Geographers have explored the complexities of everyday life, questioning meta-theories of human sexualities and gender (Oswin 2008). Location plays a key role in the formation of sexual and gendered cultures, lives and politics, and is conversely re-formed through such practices and meanings (Browne et al. 2007, Grosz 1998, Hemingway 2006). Similarly, LGBT equalities legislations are reworked through various spatial practices. Yet Hubbard (2006) and Soja (2010) have argued that cities and places are often placed in a secondary position to the social relations that they both form and are formed by. In other words, 'the city becomes a backdrop rather than an active participant in the making of new cultures and economies' (Hubbard 2006: 3). Place is important not only as a context or 'case study' but as a social relation in itself, which is often relegated below others such as class and race (Soja 2010). This contention reframes explorations of everyday lives by not only accounting for, but also taking seriously, the ways in which places diversely reconstitute lives and politics (for further discussion see Chapter 3).

In examining Brighton, a city where LGBT people were supposed to fully benefit from gender and sexualities equalities legislation, we have chosen to focus on the specificity of this urbanity. This allows us to augment well established national and international critiques of legislative change that may not account for how such changes and the state[4] have varied and diverse effects in different places, as well as differential effects within particular locales. This approach challenges

---

4   Throughout we use 'the state' to signify the manifestations of local government, police and health services, not as a homogenous entity, but one that is often read and experienced as such. We use 'statutory services' to refer to public services that must be provided by local/national government. These broad categories are used in part to protect the anonymity of participants. As we have already indicated (see Chapters 2 and 4), we see the state as heterogeneous and incoherent although it can be presented as a homogenous 'enemy'. It is the incoherence of the state that questions the oppositional politics that oppose 'it', as if 'it' is something that can be opposed.

and reworks theories that seemingly come 'from nowhere' in order to appear objective (Haraway 1991). It also seeks to address the parochialism of Western knowledge that can feign universality (Robinson 2002, 2006).

In exploring local enactments of legislative change in a specific urban context, we pursue two lines of enquiry: ordinary lives and ordinary activisms. Under the rubric of ordinary lives, we examine the messiness of social inclusions/exclusions for those who were once only stigmatised as sexual and gender dissidents. We investigate the consequences of ordinariness for those LGBT people who continue to be marginalised in the 'world we have won' (Weeks 2007). In considering ordinary activisms, we seek to explore the implications for LGBT activisms with/ within 'the state' when one is no longer positioned as a 'mis-fit', an 'enemy' or a necessary source of 'social disorder'. Ordinary activisms are understood here as activisms that work with and within potentially normalising processes and institutions and may be part of the expected and accepted order. Thus they can be in place and can be unremarkable. We will show that working with and within the state can continue to be productive, as well as regulatory and normalising and indeed, at times, unproductive. Taken together, our focus on ordinary lives and ordinary activisms holds in tension the binaries of normalisation/inclusion and radicalism/assimilation which emerge from the implementation of equalities legislation in the UK.

We use empirical data to examine the unevenly felt consequences of these new possibilities in the city of Brighton, the so-called Gay Capital of the UK (http:// www.gay.brighton.co.uk). This book draws on the rich data set developed through the participatory research project Count Me In Too which ran as an evolving project between 2005 and 2010 in Brighton (see Appendix 1 for a detailed discussion of process and methods). Count Me In Too developed organically, pursuing new lines of enquiry and alternative theoretical and conceptual developments, with LGBT people, practitioners, policy makers and academics who explored the experiences and practices of LGBT lives in Brighton. A range of stakeholders came together around the stated aim of fostering 'positive social change' (a loosely deployed term with no fixed meaning) for local LGBT people. Key actors included but also extended beyond local LGBT activists. The project provided a vehicle for interactions with LGBT people who lived, worked, socialised and used services in Brighton and Hove, to gather information about what they said that they needed and wanted (see Appendix 2 for details). This information was then explored and deployed with local service providers in ways that, stakeholders hoped, would achieve the aspirations for 'positive social change' that brought them together. The initial partnership between the University of Brighton and Spectrum (the local LGBT forum, see Chapters 2 and 6) coalesced as a research team of academics and activists,[5] nurturing the involvement of a wide range of stakeholders and developing partnerships amongst numerous and varied key informants.

5 In 2008 the research team expanded to include the University of Sussex.

During 2006, Count Me In Too surveyed LGBT lives in the city using a large-scale quantitative and qualitative survey (June–August 2006, 819 responses) and undertook a number of focus groups with identities that the project steering group named as marginalised (January–November 2006, 19 focus groups, 69 participants). Additionally, data came from feedback events and consultation events, as well as 19 interviews with 'key stakeholders' who were loosely defined as LGBT people who held some form of powerful position in the city, either in the public sector or within the context of gay Brighton (2007–2010). The data was analysed by LGBT people and service providers who then used the findings to inform policies, practice, training and grant applications. Developing beyond the practitioner based findings reports and community summaries that the project produced (available at www.countmeintoo.co.uk), this book examines this period of ostensible inclusion, experienced by some as a 'golden age',[6] to understand LGBT lives and politics. Using the findings of this project, we offer key insights into the effects and enactment of legislative change focusing on the lived experiences and everyday activisms for LGBT people.

We follow this introduction by placing *Ordinary in Brighton?* within work on urban sexualities, developing this in relation to emergent discussions of ordinary cities. This allows us to understand Brighton within discussions of both cities and sexualities and also to understand how urban gay centres can contribute to discussions of ordinary cities. We go on to introduce conceptualisations of homonormativities (a term coined by Duggan, in 2002, to name the construction of social norms that include lesbians and gay men on the condition that they conform to the norms of heteronormativity). This is a key way in which new LGBT civil rights have been understood and it begins the dialogue of our contention that discussions of normativities can be augmented by exploring the possibilities of ordinariness. Working from these crucial international debates through to the local context, we then offer an overview of the city of Brighton both as an alternative city and the gay capital of the UK, establishing the context of this book. The final section outlines the contents of this book.

Urban (Homo)Sexualities, Ordinary Cities

Urban forms are central to conceptualising the manifestations of equalities legislation. Encounters and proximities afforded by city populations and urban forms are seen as crucial to sustaining non-heterosexual possibilities, identities and practices (Bech 1997, Hubbard 2011). Cities are central to critiques of sexual and gender normalisations brought about through legislative equalities. It is timely to expand considerations of urban sexualities and genders through a consideration

---

6   Subsequently the landscape has shifted significantly, specifically with regard to the enactment of equalities legislation in an era of austerity and under a more conservative regime of the Conservative–Liberal democrat coalition which came to power in 2010.

of the 'whole city' and the experiences and practices of LGBT people within and beyond gay ghettos (Brown 2008, see also Ruting 2008, Collins 2004, Gorman-Murray 2009 and Nash 2010 for further discussion). Here we detail some key conceptualisations of cities.

Urban landscapes are formed through topographies of moral/immoral sexual and gender identities and practices. Specific spaces in cities are created either through purity or deviance in terms of sexual and gender norms (see Hubbard 2006). This has translated into territorial claims that define areas of cities, for example red light districts and gay areas such as Canal Street in Manchester, the Castro in San Francisco, Soho in London and Oxford Street in Sydney (see Hubbard 2002, Waitt and Markwell 2006, Skeggs et al. 2004, Weston 1995). These areas have acted as sites of safety, belonging and political representation. They have been a key node of investigation for sexualities geographers (see Chapter 4). However, this focus has meant that only specific areas of cities are explored and the sexualities of cities are often investigated through the consumption of 'gay scene' spaces (see Binnie 2004a, Casey 2007; Chapter 4). Cities can then be re-imagined by some through their 'gay ghettos' as safe, tolerant and accepting.

These visible gay ghettos and gentrified areas that have subsequently been labelled gay are predominantly male dominated (Podmore 1999, 2012). In part because of the focus on gay male territories in geographies of sexualities, much work has remained wedded to neighbourhood areas or ghettos 'where commercial gay venues are clustered and supported by these ['homonormative'] identities and modes of living' (Brown 2008: 1215). Following Robinson, Brown (2008) calls for investigations of the 'whole city' beyond gay ghettos (see also Ruting 2008, Collins 2004, Gorman-Murray 2009 and Nash 2010 for further discussion). Focusing on the city of Brighton, we can explore topics that move across the whole city, such as housing, safety and Pride events, rather than restricting our examination to a particular 'gay area' that is both part of, and different to, Brighton. This also enables an examination of urban governance and social movements that refuse to be mapped onto specific areas of cities. The whole city thesis has been critiqued for potentially offering a uniform conceptualisation of urban governance (see Ong 2006). We do not see urban governance as offering a blanket territorialisation or homogenous forms of control. Instead, we view the state as an assemblage of heterogeneous processes and fragmented structures (see footnote 5, Chapter 6, Andrucki and Elder 2008, Cooper 1995, Painter 2006). As Roy and AlSayyad (2006) suggest, the city is deeply fragmented with diverse governance strategies carved out for different groups.

The territorial claims that see cities labelled as gay or gay-friendly have seen certain cities being viewed as better than other (mainly rural) areas. The title of Weston's (1995) classic piece 'Get Thee to the Big City' illustrates the migratory trajectories that are presumed to enable gay life (see Chapter 3). More broadly, urban theory that has read cities in hierarchical ways has come under some critique (Brown 2008, Robinson 2002, 2006, McFarlene 2011). For example, the hierarchisation of 'world cities' and 'non-world cities' relies on conceptualisations of the urban that fail to account for the territorial locations that these urban theories arise from. For Robinson (2006) all cities are ordinary and therefore need to be explored on their own terms, not through theories developed for other cities, as all cities are 'autonomous and creative' (2).

Halberstam (2005) has challenged the assumptions that place certain cities (in the global north) as 'pinnacles' of gay life and indeed as enabling the possibilities of any LGBT existence. Cities, such as London, San Francisco, can be used to define the apexes of sexual and gender freedoms, such that a global north lens is used to judge other regions and countries (see Binnie 2004, Stella 2011). Such metronormative assumptions neglect the ways in which rural areas, towns and non-metropolitan/non-'gay' cities can enable sexual practices beyond heterosexual norms (see Kramer 1995, Gorman-Murray 2010). Halberstam (2005) uses the term metronormativities to argue that urban stories, particularly of migration from rural to urban and Global South to the 'developed' world, make assumptions about urban utopian Euro-American lives that contrast with rural lives and the closeted/'backward', 'developing' contexts. In this way, thinking about sexual/gender identities rooted in particular global north cities is used to theorise about other cities, without accounting for 'local histories, geographies and indigenous conceptualisations of homosexuality' (Brown 2008: 1225).

By exploring cities on their own terms, rather than comparing them to other cities, we can highlight and discuss the formations and effects of metronormativities. In this book, we seek to make the metronormative strange, by taking it as our object of investigation. We do so by exploring the specificity of Brighton, seeking to shed light on the ways in which lives and politics are built in part by place.

Working across the whole city and addressing it on its own terms does not necessarily negate networks and flows between local, national and global scales (see also Binnie 2004, Massey 2005). Smith (2005) demonstrates the multi-scalarity of LGBT citizenship activisms, illustrating the particularity of certain urban contexts and how queer politics is differentially produced at different scales. Following this, we see networks, flows and relations between scales as influencing and creating the 'specific city-ness' of each city. Alongside critiques of national/supra-national equalities policies and practices (see for example Binnie 2004, Stychin 2002) local city-ness needs attending to in considering the effects of equalities legislation. Such a focus offers insights into the reconstitution of everyday lives and activisms. However, as we now move to consider, scholarly examinations of (homo)normative social orders can be augmented with an attention to the possibilities, as well as the normalisation, of state and cultural inclusions.

Homonormativities

Homonormativity has offered important insights into how once deviant forms of sexuality become acceptable, in ways that are premised on the normativity of seeking certain forms of 'equal rights', rather than sexual and gender liberation

(Duggan 2002, O'Brien 2008; Richardson 2004, 2005). Work has been undertaken in this vein to explore how power relations and social and political hierarchies crosscut sexual/gender deviancy, creating both privilege and marginalisation (Bryant 2008, Lim and Browne 2009, Hines 2010, Vidal-Ortiz 2008). This body of material illustrates how homonormalisations are differentially accessed in gendered, classed, racialised, aged and sexualised ways (see for example Hines 2007, Seidman 2002, Taylor et al. 2011, Taylor 2007, 2011). Normalisations are therefore exclusionary (see Appendix 1 for a discussion of intersectionalities and multiple marginalisations). Only certain types of sexual 'deviance' are rendered 'equal now'.[7] Not all who come under the category lesbian, gay, bisexual and trans can be, or indeed wish to be 'normal'. This can be particularly painful for those who believe that they now 'should' be included (see Chapters 3 and 5).

These new (homo)normalisations are viewed as regulating and restricting citizenship gains to particular (homo)normative regimes (see for example Duggan 2002, Sears 2005, Stychin 2003, Warner 1999; Richardson 2005). Alongside acceptable heterosexualities, sexual citizens are thus increasingly defined through the paradigm of culturally acceptable homosexuality, or the 'good gay' citizen (Bell and Binnie 2000, Seidman 2002, Stychin 2003). 'Good gay' citizens are seen to fit into particular heteronormative orders pertaining to monogamy, domesticity, capitalist individualism, consumerism as well as class and racialised positioning (see Evans 1993, Warner 1993). This can also apply to trans rights and may also create good/bad trans citizens (Monro 2003, 2005, Valentine 2007). Therefore, critical scholars have pointed to LGBT inclusion, diversity strategies and equalities legislation as drivers that develop normalising liberal citizenships.

Cooper (2006) argues that the liberal reading of sexualities equalities saw it as analogous with race and gender equalities, and thus failed to question heterosexuality or seek to deconstruct the privileges of sexual orientation (just as whiteness and masculine privilege remained intact). Indeed discourses of challenging homophobia/sexual 'liberation' can be used to perpetuate discrimination and othering, for example stereotyping particular groups as being homophobic and deploying this to justify engaging in conflicts such as the war in Afghanistan (see for example, Bryant 2008, Puar 2007).[8]

---

7 For example, during this period there was a 'crack down' on 'extreme porn' (see Attwood 2010, 2011, Wilkinson 2011), and sex work (see Campbell and O'Neill 2006, Hubbard 2011). Much of this was framed in discourses of violence against, and trafficking of, women and children, which sought equalities agenda based on 'sexism'. This created new sexual 'others'/'dissidents', and shows the complexities of sexual/gendered equalities legislations, where certain understandings of 'sexism' and 'homophobia' can be deployed.

8 Homonormativity is read across multiple scales, from the domestic sphere of monogamous, domestic partnerships to the nation state. Homonationalism, a term coined by Puar (2007) seeks to explicate how national identities, military actions and other state supported violence can use a 'tolerance' of homosexuality to construct nations such as the UK and the USA as 'better than' other countries, justifying violence, military 'interventions' and discrimination through supposedly 'improving' human rights. It develops from

A key way in which urban gay (and to a lesser extent lesbian) lives have been celebrated has been in relation to the economic benefits that 'the creative classes' bring, which are particularly epitomised by gay men (Florida 2004). In the 'new urban politics', cities compete for 'creative classes', seen as the bearers of wealth and prestige, with clustering of activities into particular areas (such as 'cultural quarters') (see Bell and Jayne 2004). 'Cosmopolitan', 'modern' and 'cutting edge' cities become signified through the presence of large gay areas and commercial scenes (Bell and Binnie 2000, Binnie and Skeggs 2004, Knopp 1990, 1998, Ruting 2008, Skeggs et al. 2004). The move to 'post-industrial' consuming cities has had a significant impact on such sexual politics, with certain cities being sold as cosmopolitan through their 'gay quarter' (Binnie 2004a, b), and there have been losses as a result of the sanitisation of gay city zones for 'good' gay consumption (Delany 2001). Added to this, changes in legislative citizenships, particularly in the Global North, have led Brown (2008: 1224) to argue that we need new vocabularies to explore the spatial organisations of cities across 'the whole of urban space', being attentive to histories and geographies that are specific to each city, recognising that 'these spaces are "ordinary spaces" in "ordinary cities"'.

Undoubtedly the celebration of gay identities as part of creating cosmopolitan cities has privileged some subjects, particularly certain gay men. In contrast to earlier scholarly explorations of gay ghettos as the locations of political– economic resistances to heteronormative orders (see for example Lauria and Knopp 1995), current scholarship largely reads these as expressions of neo-liberalist tendencies and purported bastions of 'gay' tolerance. For example, a key point of contention has been the marketing of cities as locations of diversity, creativity and cosmopolitanism, which is evidenced, in part through the supposed 'acceptances' of gay (and lesbian) people, regardless of whether they have national legal rights. Authors have critiqued the deployment of gay marketing in regeneration processes and drawn attention to the fissures and gaps of such acceptances, including those identified above (see Doan 2008, Halberstam 2005, Hughes 2003, Rushbrook 2002).

Queer thinking has sought to challenge sexual and gender normativities as produced through liberal equalities gains, working towards non-normative forms of

---

understandings of the heterosexualisation of 'the state' through legislation and directives and can be used to understand how states can normalise and incorporate particular forms of 'good gay' citizens through, for example, serving in the military.

In addition, homonormativity does not only relate to sexuality. Bryant (2008) notes that homonormativity was in use by trans-activists since the 1990s. Homonormativities could then be understood as the hierarchisations of trans identities as well as within the relations between LGBTQ people. Much more work has been done on the former with Doan (2007), for example, arguing that what she terms 'queer spaces' are theorised as providing a safe haven for LGBT oppressed by heteronormativity of most urban areas. However, queer spaces are themselves gendered and so provide only limited protection from harassment and violence for trans people (see also Browne and Lim, 2010; Nash, 2010; Lim and Browne, 2009).

liberation. Cooper (2004: 142) explores 'the question of how to create and sustain ways of being and doing that counter, thwart or trouble dominant social relations'. This focus has been a key way of considering LGBT and particularly queer modes of engagement with new equalities landscapes. Queer anti-normativities are often counter-posed against liberal bureaucracies. When considering activisms, assimilation is placed in opposition to radical reconsiderations. The emerging framework of civil rights that incorporates, rather than demonises, LGBT people is viewed by some as having led to deradicalisation, such that sexual and gender liberations have been replaced by equalities legislation and access to conservatising institutions (Duggan 2002, Richardson 2004, 2005, Sears 2005). Thus ordinariness can be read through assimilation, normalisation and conformity, which no longer offers resistance or transgression to gender and sexual inequalities (see Brown 2008, Ward 2008).

Ordinariness

Scholarship which has examined homonormativities and undertaken important work in highlighting normalisations and exclusions may not account for all the possibilities of ordinariness. Weeks (2007: 9) contends we should 'never underestimate the importance of being ordinary'. This book begins by separating ordinariness and normativities. Ordinariness is related to 'normalisations', but we also will seek to argue that it can offer more than the regulatory framing that normalisations can imply. We do this by examining what happened when certain sexual and gender dissidents supposedly became ordinary, rather than exceptional and transgressive. We will argue that ordinary lives can be desirable and do not necessarily require adherence to (hetero)normativities. Thus, rather than looking at the cultural concept of 'ordinariness', we are interested in the move to being ordinary as enabling LGBT people to be in place, unremarkable and an accepted part of urban lives and politics. This offers insights into the multiple possibilities and hopes of becoming ordinary, as well as the pain of exclusions and otherings from new normalities.

In seeking to explore more than the limitations of the inclusions of LGBT people in the gay city, we draw on theorising that has queried the focus only on the critical, that lead to hopelessness. Sedgwick argues that, in contemporary critical (social) theorising, to work 'out of anything but a paranoid critical stance has come to seem naïve, pious and complacent' (2003: 126). She argues that we constantly look for evidence of systematic oppression and that this is seemingly ubiquitous. Similarly, Koch and Latham (2012: 516) contend that forms of urban scholarship have 'highly developed skills in diagnostic critique, but are often not very good at knowing what *does* work or *how*'. Here we seek to examine what did work without negating the charges of collusion and normalisation and the problematic ways things work at the expense of other possibilities.

Cooper's (2009: 105) term 'alternormativity' is useful in starting this journey. She uses this concept to enable an understanding of 'organised social practices that neither replicate nor straightforwardly reverse hegemonic relations'. This conceptualisation can work beyond the binaries of hetero/homonormativity and resistance/mainstream, and recognises that relations of power continue to need critical exploration. It also affords the possibility of reworking the hegemonic sexual and gender norms from within institutions such as local government. The events that Cooper investigates do 'not simply reverse the world outside but evolve to sit at a far more complex angle to it' (Cooper 2009, 126). This enables complex conceptualisations of social practices and the activisms that seek to (re) work them, including those that are simultaneously complicit and resistant, inside and outside (Rayside 1988, see Chapters 6, 7 and 8).

We begin from the premise that LGBT lives and activisms can 'sit at a complex angle' to power relations enacted by state actors, powerful (possibly homonormative) LGBT people and others who deploy resources to (re)produce inequitable relations. This enables a multifaceted examination of the effects of equalities legislations and the material manifestations of inclusions and marginalisations. Such a discussion does not negate regulation, disciplining of bodies and normalisations whilst *also* allowing for an exploration of the possibilities of ordinariness.

Brighton and Hove: The Extra-ordinary, Ordinary City

> Brighton is on the edge ... the gay capital of the South, the location of the dirty weekend, has historically embodied the genitals rather than the heart. Its sexual ambiguity is present on the street, in its architecture from the orbicular tits of the King George's Pavilion domes, to the giant plastic dancer's legs which extrude invitingly above the entrance of the alternative cinema ... its sexual history is a memory cathecting contemporary erotic identifications as erotic, decadent, degenerative and whorelike. (Munt 1995, 104)

In this section we return to emphasise the import of place as key to understanding that which might become ordinary. Brighton, as the focus of this book, offers a city context which affords an investigation of how the specific city-ness of a city can play a part in the reconstitution of gender and sexual lives. In order to take Brighton on its own terms, we seek to offer the reader an introduction to Brighton. Consequently, in this section we begin by broadly outlining key characteristics and imaginings of this extra-ordinary/alternative city and then explore the specific sense of place associated with gay Brighton. However, a city can never be contained in the pages of a book. We have chosen to focus on specific features of this city in order to illuminate aspects of it that are relevant to discussions in this volume.

The City of Brighton and Hove, located in the affluent South-East region of the UK, is generally recognised as having experienced rapid economic growth in the late twentieth and early twenty-first centuries. Characterised by service sector industries, including IT, new media, tourism and hospitality, and its proximity to London (an hour by train allows for commuting to the capital, see below), the city attracts high wage professionals. However, it also attracts a significant percentage of people in lower income categories (almost a third of the workforce earned less than £250 a week, see May 2003). The city's population is recorded as predominantly White British (89.5 per cent) and is read by some as celebrating a middle class diversity whilst vilifying working class identities and practices (see Burchill and Raven 2007; Chapter 4). The majority of the economy is based on tourism and retail, 'the purpose is not production, it is consumption' (Gough 2005: 93, Collis 2010). This is coupled with a booming housing market with house prices rising rapidly from the 1990's and purportedly recovering from a dip in 2009 to hit record highs of over a 16 per cent increase in 2010 compared to the previous year (see Thompson 2010, http://www.housepricesuk.org.uk/property/brighton-hove. htm, see Chapters 3 and 5).

In the city branding of Brighton and Hove, 'Brighton' is sold as 'vibrant, colourful and creative' (www.visitbrighton.com, 2011). Thus, it is important to make the distinction between the mythologies of 'Brighton' and the 'city of Brighton and Hove' (which was established in 2000 after Brighton joined with Hove and the neighbouring district of Portslade as a single authority in 1997). 'Brighton' is an imagined place and, when speaking of 'Brighton' in this book, we talk about this geographical imaginary, marketing and perceptions that may or may not relate to the boundaries of the city. In contrast, Brighton and Hove is a political, unitary authority with clear boundaries and this term relates in particular to the governance of this region and services provided therein. Whilst there is much overlap between Brighton/Brighton and Hove, there are also distinctions that are sufficiently significant that for the remainder of this book, we use Brighton when speaking of the imaginings, marketing and perceptions of this place, and Brighton and Hove when referring directly to public sector/statutory services, political representation and other formal measures, as well as the specificities of the boundaried city.[9]

The 'liberal' community of Brighton, with its bohemian pretences, middle class values and plethora of organic and 'green' shops is described as the 'vanguard of cultural revolution' (Burchill and Raven 2007). Brighton is often imagined as a place of radical politics, which was to some extent borne out in the election of the first Green Party Member of Parliament, Caroline Lucas, 2010, in a Brighton constituency. This was seen to indicate that Brighton is 'as a kind of British

---

9   It should be noted that many of the participants in the CMIT research who are employed in services across the city discussed 'Brighton and Hove', whereas others spoke of 'Brighton'. This may be attributed to the corporatisation of language, as well as the formal interview settings. Where they use 'Brighton and Hove', we respect this, as the overlap in boundaries and 'official' boundaries and imaginings are clear.

equivalent of San Francisco: a bohemian city with a massive gay population and a penchant for alternative free-thinking' (Petridis 2010).

Brighton has a long history of association with and relatedness to London (Thomas 2011). 'Lubricated by its proximity to London' (Hemingway 2006: 437), contemporary Brighton as 'London by the sea' (Tomlinson 2005) continues to be used as an escape from the capital city:

> [Brighton] occupies a unique place in London's imaginary, associated with freedom from constraint, the non-official and the inside-out; a place not just of leisure but resistance to power; of celebration of the body and desire. (Thomas 2011: 4)

Pleasure is important in these imaginings: as Gough (2005: 88) contends, it is 'a city dedicated to pleasure all the time'. Period architecture has been invoked as a key facet of this (classed) pleasure,[10] such that:

> [Brighton] is the only urban environment in Britain that could be called 'delirious'. Builders sought to express an emotional response to place, to confect feeling out of the inanimate stuff of construction, to make the intrinsically static into exuberant artifice; buildings that appear to sway in the breeze as they bask in the sun. (Gough 2005: 93)

Brighton's intersecting characteristics of the city and the sea makes it a populous seaside resort. It has a long history of tourism related to its seaside location. The seaside is not 'an inert or incidental surface variation but a landscape that connotes social, moral and cultural values and meanings which are territorialised' (Hemmingway 2006: 432). As a ludic seaside town, Brighton is a place to 'play' in with multiple 'essences', as Campbell (2002: 12) notes about the Brighton Museum:

> Here different Brightons – vulgar postcard Brighton (Carry on at Your Convenience), louche Brighton (The First Gentleman, The Gay Divorcee), the Brighton of the disaffected young (Quadrophenia) and the Brighton that mixes violence and fairground glitter (Brighton Rock, Oh! What a Lovely War) – are extracted as simple essences from the mix that is the town itself.

The celebrated resistances of Brighton include histories of youth transgressions, for example the Mods and Rockers (see Cohen 2002) and the portrayal of anarchic pleasure in films such as Carry on at Your Convenience (1971) (Thomas 2011).

---

10   For Gough, recent modern building 'egalitarian for everyman' fails to achieve the 'licentiousness and pleasure' of earlier buildings, and is instead 'the very personification of the not-for-pleasure principle' (2005: 95). This symbolically and materially illustrates the classed based nature of the pleasure of Brighton (see also Chapter 5).

Brighton is a liminal seaside town and a tourist space where 'normal' routines can be suspended so that escape/pleasure/hedonism/excess or at least 'other' experiences become possible. It exists on 'the edge' figuratively, as well as literally through a beach and a pier that exists between land and sea (see for example Clisby 2009, Shields 1990, Munt 1995, Preston-Whyte 2008, Webb 2005, Tomlinson 2005). This 'edge' can be seen in socio-cultural expectations of the seaside, where 'there are long associations of the seaside with eroticism, transgression and the carnivalesque' (Clisby 2009: 53).

(Hetero)sexual transgressions, 'of the body and desire' are important to understanding the 'edge' which characterises the:

> liminal sphere of the seaside: the hedonistic transgressive sexual mores of the holiday resort as a place where sexual desire can be pursued outside of the normal constraints of everyday life. (Clisby 2009: 48)

Piers, seaside promenades and other features of the seaside resort are seen as providing the possibilities of sexual encounters and liaisons (Clisby 2009, Tomlinson 2005). A common imagining of Brighton is one based on (hetero) sexuality and a history of the 'dirty weekend' of heterosexual intra/extra-marital liaisons and (hetero)sexual indulgence (Hemmingway 2006, Munt 1995, Shields 1990, 1991, Thomas 2011).

However, the heterosexual (classed) excesses of seaside Brighton are often relegated to 'the past' (literally memorialised in the Brighton museum) in favour of urban chic. Seen as 'seedy' and 'vulgar', Brighton's tourist industry in the twenty-first century has sought to distance itself from the 'kiss me quick' and its reputation as the epitome of 'fighting, sex, misdemenour – the pure pavilion of transgression' (Jenks 2003: 162), as well as the 'drug death capital' (Ryan 2010). Appealing to the modern sophisticated traveller, who partakes in the lucrative conference tourism industry and cosmopolitan 'weekend breaks',[11] the image of Brighton has sought to tame the excesses of (sexual/classed) deviance (see Clisby 2009, Hemingway 2006).[12] This is epitomised through branding of 'the City by the Sea'[13] (May 2003) seeking a 'sophisticated' seaside city image that elides its seaside resort past, including through links to the arts (for example the Brighton Festival). Yet, in order for some to experience escape, freedom, relaxation, rest

---

11   Further work is needed in this area both to consider the interactions of the city and the sea and the ways that certain imaginings relegate others to the 'pages of history'.

12   Linked to teenage pregnancy, disease and other vices, holiday sex, particularly for young working class women, is understood as 'transgressive, deviant and problematic behaviour' (Clisby 2009: 48). Clisby notes that the moral panics surrounding working class sexual excesses, the consequences for 'the taxpayer' and the 'kinds of bodies that are reproductively authorized' (52) serve to (re)place specific social hierarchies.

13   A marketing phrase coined by the council that is widely used both inside and outside of the city (www.thisbrighton.co.uk/bitspieces.htm).

and so on, 'others' are cast as unwelcome and undesirable (Preston-Whyte 2008: 352). Interestingly the gay capital image has been part of the sophisticated, modern Brighton presented in the first decade of the 21st century.

Gay Brighton

Brighton is a city with a long history of sexual transgressions and prides itself on 'leading the way' in terms of LGBT equalities agendas. Despite being the self-appointed 'gay capital of the UK' (http://www.gay.brighton.co.uk), it is a small city (with a population of approximately 250,000)[14] and as such exists between the 'large' urban gay metropolis discussed in the literature and the small 'ruralities' that are often read as heteronormative (see Brown and Knopp 2003). Challenging how cities have been seen in terms of urbanism, large cities can be insular and parochial and small cities, particularly those with networks and connections like Brighton, can be read as 'cosmopolitan' in part through their 'acceptances' of gay populations (see Brown 2008).

It is claimed that 'Brighton's gay history is as old as the town itself' (Collis 2010: 133). The narration of lesbian, gay and bisexual histories (note, trans people are not included in the local 'Ourstory' project – www.brightonourstory.co.uk) since the 1800s, has been used to place Brighton as a town where gay lesbian and bisexual sexual contact, arrests and bars have an extended history and thus also lay claim on the area (see www.brightonourstory.co.uk). Indeed Brighton's coastal location has been seen as a key factor.

Brighton in the first decade of the twenty-first century was sold as an 'escape' from the 'repression' of everyday life, a place to 'be yourself', and this is crucial for gay tourism (see www.brightonourstory.co.uk). There was a move at the turn of the century towards attracting national and international gay tourism. The official sanctioning of this 'transgressive' behaviour took hold as the move towards legislative equalities was also taking place. Because of these changes, the tourist drives were able to feed into and from the impulses to 'clean up' Brighton's image whilst maintaining the 'edginess' and 'trendiness' of the city (see also Binnie 2004a, Delany 2001, Florida 2004).

As part of its tourism promotion, the local government website sold gay Brighton in relation to its clubs, bars, shops, saunas, beaches and services (including estate agents, local voluntary groups and the police). Gay listings represented a Brighton that parties every night with 'something for everyone'. Indeed much of Brighton's history is 'linked to its gay pubs and clubs' (Collis 2010: 134), reflecting the pleasurable consumption of the city. Brighton, in contrast to other cities with notable gay scenes, has a relatively (and unusually) large

---

14   It ranks 44th in the UK in terms of the district's population size (see http://en.wikipedia.org/wiki/List_of_English_districts_by_population) and has a population of approximately 250,000.

number of scene venues for its size and LGBT community spaces and activities coalesced around the sale and consumption of alcohol and entertainment. Figure 1.1, which locates businesses in Brighton that paid for advertising in *Gscene* (a local LGBT/gay magazine), also shows the clustering of these businesses into particular areas. The expanded box in Figure 1.1 shows the Kemptown area, often portrayed as 'the gay village'. However, as we explore in Chapter 3, there was more to 'gay Brighton' than just these venues. The 'scene' extended beyond the boundaries of this area and the situation of such businesses questions the location of the scene in Brighton in the 'gay ghetto'.[15] The notion of 'the scene' in this research predominantly referred to 'gay' scenes, rather than LGBT spaces (see Chapter 4). Of the businesses shown on the map in Figure 1.1, two were defined as 'lesbian/women's bars' and one other provided a weekly lesbian/women's targeted night (Girls On Top).[16] Consequently we use the term 'gay scene', but discuss the spectrum of LGBT experiences with and on 'the gay scene', rather than focusing on broader non-commercial (potentially more 'radical') LGBT/queer social spaces.[17] Closely aligned with the scene, the city is also promoted by its large annual Pride event which happens usually on the first weekend of August (see Chapter 8).

The presence of LGBT people in urban areas has been demonstrated through the existence of marked gay districts. In Brighton estimated population figures in the whole city are also often invoked to lay claim to the 'gay city' moniker. In contrast to the United States where political power has been part of the realisation of 'gay ghettos', gay areas in the UK are not linked to politics in this way (see Knopp 2008). However, in Brighton the 'gay vote' is courted and key figures buy into and reiterate the imagining of an LGBT 'community'. For example, in her welcome address at the Transforming LGBT Lives conference (2010), Mary Mears (Deputy Leader of the Council 2010) spoke of an LGBT 'community' that brings business and tourism to the city, and Caroline Lucas has spoken of being 'proud' of Brighton's LGBT community. In this way, 'the LGBT community' is

---

15   Indeed Wild Fruit takes place on a Sunday night in West Street, a location that is perceived to be hyper-heterosexual.

16   The only women's nightclub (Candy Bar) shut its doors in 2009, naming a lack of support from women as a key reason that the business did not succeed. Defining any of the current venues as 'lesbian' is difficult. Whilst most venues (claim to) have a mix of clientele – mainly cisgendered male/female, although there is a large drag queen circuit – the 'lesbian' bars are managed by women and known to be 'lesbian'. In contrast to some bars and clubs that continue to operate door policies that encourage a mainly male client base, these bars do not. In the main, most bars/clubs that make up the scene could be defined as 'mixed' in terms of gender. Further work is needed on the histories and development of these spaces similar to Nash (2006) who explores the histories of lesbian bars which have in the main been subsumed to 'queer', mixed spaces.

17   These have increased in importance and political influence since this research was undertaken, and are unfortunately beyond the scope of this book. Further work is needed to explore the events, networks, groups and activities that could come under the political as well as the umbrella term 'queer'.

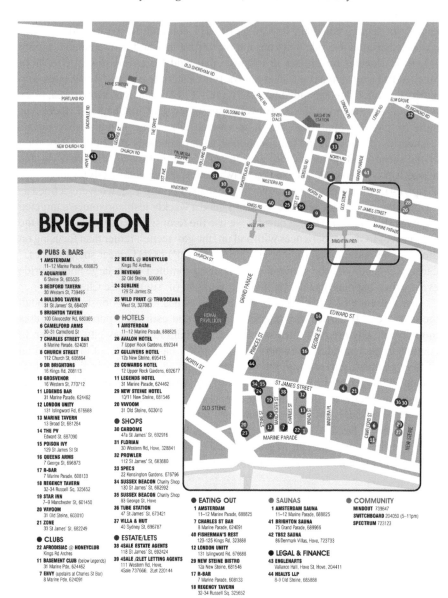

Figure 1.1 Map of 'the Brighton scene'

Source: Printed by kind permission of Gscene Magazine Ltd.

created as a 'unique' element of the city. This is in part due to the estimated size of the LGBT population, which is impossible to measure with any certainty.[18] Public Health (see Memon and Walker 2008), local government (BHCC 2008) and other documentation assert that 'Brighton and Hove has the largest concentration of lesbian, gay, bisexual and transgender people in England outside London' (Memon and Walker 2008: 32), placing this figure at 10–15 per cent of the population, or around 35,000 people[19] (see also Collis 2010). These imaginings have effects.

The heterosexualised 'dirty weekend' in Brighton references more than extra-marital affairs. As late as 1997, health authorities claimed 'there is no local gay community' in order to claim there was not a specific HIV issue in the city (cited in Collis 2010: 2). However, after 1997 the presence of a supposedly large LGBT community, in the light of New Labour imperatives to consult with local communities, offered some leverage in mainstream political processes and the provision of services for LGBT people (see Chapters 3, 4, 7 and 8). In addition to this, there is the legal requirement to provide services that cater for all sectors of the population. Thus, in the period that this book covers (2000–2010), statutory services such as the council, police and primary care trust gave financial support to organisations such as Pride in Brighton and Hove, Spectrum and other local community groups for LGBT people, as well as seeking to work with such organisations. Moreover, the sexual and gender equalities policies and initiatives developed at a national level were (ostensibly) not contested or opposed by local government or public sector services (such as police and health services) here.

---

18   Not only because the census 'doesn't count' this, but also because the category itself is unstable, and not all who might be recognised within this grouping identify with this category. Thus, 'accurate measurements' rely on quantifiable 'known' categories in boxes that are easily 'ticked', and this is problematic for LGBT people (see Browne 2010b). This was shown most vividly in 2010, when the Office for National Statistics unveiled data from its Integrated Household Survey, placing the LGB (*sic*) population at 1.5 per cent or just 726,000 (Joloza et al. 2010). This research placed London as having the 'highest concentration' of LGB people (2.2 per cent), but the findings did not break down areas of the South-East (1.3 per cent). The estimated LGBT figures for Brighton and Hove challenge the assertion that London has the 'highest concentration' of LGB people (Joloza et al. 2010).

Nationally, the absence of figures had led for a call for census and other forms of data that can measure sexual orientation (the drive for data on trans people has been separated from this call, due to the location of trans issues within gender equality). This is believed to enable a visible LGB population that will then necessitate service provision and therefore further the equalities agenda. Whilst there are many ways that such a call can be contested (for example does visibility mean equality? (see Tucker 2009); what does identifying as LGB mean?; what are the dangers of this?), this call has in part subsided following the recent testing of a 'sexual identities' question, which returned a figure of less than two per cent of LGB people (see Browne 2010b).

19   The only available, contentious, figures are those that suggest that over 2 per cent of couples in Brighton & Hove are same-sex (Duncan and Smith 2006), and these measurements of same sex couples are challenged (see also Brown 2000).

Brighton is a city in which LGBT people were seen to 'become ordinary' in the New Labour Era (1997–2010) in the UK. Brighton was thus recognised nationally and internationally as a place where LGBT (or at least gay) identity politics are supposed to have been 'successful' in establishing a 'community' that is heard and listened to under the banner 'LGBT'.

Brighton as the 'gay capital of the UK' is often viewed as an exceptional space (see for example Collis 2010, Munt 1995, Gough 2005), a city that should be setting an example for others in terms of sexual and gender liberation. Despite this Thomas (2011: 4) argues: 'Brighton as it is represented in literature and cinema rarely features its ordinary residents, making a living out of the tourists or commuting to London'. Furthermore, 'there is something of a hole where the gay literature of Brighton should be' (Thomas 2011: 4, however see Munt 1995). This book hopes to excite and perhaps entice responses that address these lacunae.

In many ways, Gay Brighton represents the stereotype of a Global North city dominated by commercial leisure venues that have emerged from secrecy and flourished in an era of lesbian and gay acceptance. We could attempt to universalise this city offering 'lessons' for other places. Yet, Robinson (2005) argues that there is a need to parochialise Western knowledge, as we query the ubiquitousness of 'learning' derived from world cities. Many of the discussions of homonormativities are written from, or draw on, perspectives developed in North America (Brown, 2012). Too often there has been a tendency to assume that this North American perspective holds universally. Homonormativities have thus been found across the world without investigating the contexts of these normalisations. This can mean that geographical specificities are elided. For example, the homonormativities of the UK under the national governance of New Labour had features that contrasted starkly with the USA under the Bush (Republican, right-wing) political context. Whilst homonormativities are important in the British, or more precisely English, context that we explore, our focus on the local and specific city-ness of Brighton illustrates the differentiations of how legislative, social and economic contexts come into being in ways that are specific to a particular place. In attending to the possibilities of ordinariness in Brighton, alongside the exclusionary normalisations, we respond to the call for social relations to be located and for that location to be taken seriously.

Outline of the Book

Having now provided the key orientations of *Ordinary in Brighton?*, the next chapter offers some insights into the methodologies of the Count Me In Too[20]

---

20 Appendix 2 associated with this chapter reviews the specific methods of undertaking research with minority and marginalised communities in ways that can generate findings of relevance to these communities and a range of service providers in the public and third sector.

research project. Exploring the context of the research and how its 'successes' are based in shifting contexts and expectations, the chapter argues that place, in this case the city of Brighton, needs to be attended to in examinations of participatory and other forms of research practice.

Chapter 3 initiates a discussion of *Ordinary Lives*, the theme of the first part of this book, by examining the social and political implications of imagining Brighton as the gay capital of the UK in the first decade of the twenty-first century. The chapter illustrates the importance of urban imaginaries in the constitution of LGBT lives, even in national contexts where LGBT equalities seemingly 'apply everywhere' (Monro 2010). We examine whether gay Brighton in the early twenty-first century lived up to the utopic visions of urban (sexual and gendered) lives and freedoms for all (see Aldrich 2004, Bech 1997). We explore the contestations of this label where LGBT people did not experience the promises of migrating to, and living in, this city. City imaginings are investigated in terms of their productive possibilities of failure (see also Halberstam 2011). The city (incorporating public services and scene spaces) was supposed to be better. The imperative implied in this '*supposed* to' created a place-based impetus for LGBT activisms that recreate LGBT lives and act as a catalyst for LGBT politics.

Chapter 4 examines the 'gay scene' in Brighton as a key aspect that defines gay cities, as well as the focus of discussions of homonormative spaces. We augment discussions of intersectionality and multiple marginalisations (see Appendix 1) from supposedly inclusive spaces, on the basis of two social identities: lesbian and class. We then discuss a form of othering that unexpectedly appeared in the research: not drinking can be a form of exclusion. The chapter then moves to examine key debates regarding the commercialisation of gay scenes and how this relates to a supposedly non-commercial community imperative.

Chapter 5 brings out the differential experiences of mainstream and LGBT service provision, discussing a different manifestation of multiple marginalisation and the contestations of the utopic visions of the gay capital. In putting bi lives and trans lives into conversation, we contend that gay Brighton excluded bi people and trans people in overlapping ways. In this way we show the pervasiveness of heteronormativities and how these are experienced differently within LGBT collectives. The materialities of exclusion were painful and at times life-threatening. However, and building from the productive failures of Chapter 3, in the expectation that this city will cater for 'all' LGBT people, there were also possibilities. Various forms of trans activisms and bi activisms were made available as a result of the expectations of how equalities landscapes should be enacted in gay Brighton.

The second part of the book, beginning with Chapter 6 explores some of the ways in which the imperative to be the gay capital of the UK was manifest. We argue that LGBT work was an accepted part of local government and other state provided services, contesting the dichotomy of radical/assimilationist politics. The new legislative era of the early twenty-first century offered possibilities, as well as normalisations and recuperations. The activisms we discuss exist in very different contexts to those identified in the 1990s (which sought to change legislation or work against heteronormativities in institutional spaces; Cooper 1994, 1995). These activisms worked with and within the state and can be dismissed as 'selling out' in neoliberal eras. However, we argue for the possibilities of politics with/ within new equalities landscapes. We use an in-depth examination of Spectrum, an LGBT community and voluntary sector group, to explore the possibilities and costs of working with the state.

Chapter 7 augments understandings of the nuances of 'partnership working' between the state and LGBT groups and individuals. Contesting the assumption that dealing with legally defined 'hate crime' makes people safe, we demonstrate the disparate ways in which safety was enacted and partnerships were operationalised and contested. By exploring critically the attempts to create safety through partnership working, we show that resistances and working together are not mutually exclusive but rather that these tactics were variously deployed.

Chapter 8 offers a different perspective on the politics of Pride, including an exploration of the possibilities of celebration, positive belonging and community, without negating nostalgia for an earlier era when political 'sides' were clearly divided. Challenging the notion of protest as the only form of political action possible during Pride events, we show (the limited, flawed and commercial ways) in which Pride in Brighton and Hove as an organisation sought to provide space for LGBT people to represent themselves by refusing singular homonormative agendas. We finish by contesting the bracketing of certain lesbians and gay men as homonormative, by pointing to the deployment of power by certain gay men in ways that targeted women. In this way we move the politics of Pride beyond a discussion of a one-day event to also account for the ways in which these festivals and organisations shape and are shaped by cities throughout the year.

Chapter 9 acts as a concluding chapter drawing together key arguments regarding ordinariness. We explore the possibilities of ordinariness as creating the potential for inclusion, and contend that, unlike normalisation, ordinariness does not require ideals nor is it necessarily exclusionary. We use this to advocate for a politics of ordinariness that sits alongside the desires to be queer in ways that contest normativities, but move from ordinariness to introduce a conceptualisation of commonplaceness. Finally, we finish the book by seeking to inform future research that could deal with questions that are emerging locally, nationally and internationally post-2010.

Chapter 2

Contextualising Research and Researching Contexts: Situating Participatory Projects

Introduction

> Emphasis on context, process, social interaction, material practices, ambiguity, disagreement – in short the frequently idiosyncratic and always performative nature of learning – [is] being lost to formulaic distillations ... and instrumentalist applications. (Amin and Roberts 2007: 353)

Participatory research has become an important facet of social research methodologies. This book draws on the research of the Count Me In Too project, which is characterised as 'participatory', contesting hierarchies and working collaboratively (see for example, Cahill 2007, Cornwall and Jewkes 1995, Gatenby and Humphries 2000, Kindon et al. 2008, Pain 2004, mrs kinpaisby 2008, Pain and Francis 2003, Stoeker 1999, Ward 2007). Such a participatory methodology fosters approaches that seek to be inclusive and that seek to re-position communities, activists, service users and others as not simply subjects of research or recipients of services, but as 'central to the solution of social problems' (Taylor 1999: 372). Yet, participatory approaches do not circumvent power, and participatory research is not inherently or necessarily progressive (Kapoor 2005, Kesby 2007, Pain 2004, Taylor 1999). Nevertheless, participatory spaces, such as that produced through the processes of Count Me In Too, can open up socio-spatial arenas that differ from everyday lives, in ways that 'valorise equality, encourage self-efficacy' (Kesby 2007: 2819). Empowerment in participatory research can thus be understood as power *with* others, rather than an absence of power, erasure of power differentials or achieving power over others (Kesby 2007, see also Allen 2003).

Whilst much has been written about the processes and outcomes of participatory research, we believe that it is also important to reflect on the latent potentials and conditions that may be needed in order for participatory research (and partnership working, see Chapter 6) to attain the ideals that are promised, that is, positive social change and inclusion. Focusing on finding rules, process and practices can transform research 'into a package – discrete and manageable to suit the institutional culture, and modular to make it flexible and transferable to various project sizes, tasks and contexts' (Kapoor 2005: 1211). This neglects the ways in which various spatial forms recreate knowing and practice by means that cannot be coded and generalised (Amin and Roberts 2007, and above). We contend here that research should be considered within its territorial and temporal context and in this way be judged on its own terms by those involved in and affected by it. This approach questions meta-theoretical, positivistic impulses of research design and data collection.

This chapter begins from the assertion that context needs to be attended to as much more than a backdrop in examinations of participatory research practice, as well as other forms of research practice. It examines the antecedents of the Count Me In Too (CMIT) research and explores how its 'successes' are based in shifting contexts and expectations.[1] We begin by considering the possibility that participatory research can only effect social change 'when the appropriate conditions prevail' (Taylor 1999: 376). We deploy a descriptive approach to question the assumption of ubiquitous possibilities of participatory research, and highlight the importance of reflecting on social, cultural, economic and political situational specificities that create research. Using this framework, we then critically reflect on the success of CMIT in 'effecting positive social change for LGBT people', contending that successful research also needs to be explored in its territorial context.

Researching Context: What Makes Participatory Research Possible and Useful?

> Meg: Questions arise like 'Is there on-going value in relationships between academia and activism?' and my own personal opinion is 'Yes'. But what's the balance and where do you take it forward? It depends on the topology of our community as to what academic data would be useful in that space. (Feedback group, November, 2010)[2]

While some accounts of participatory research speak of researchers reaching out to 'disenfranchised' people who need 'empowering', CMIT began in temporal and spatial context where there were LGBT people who already felt empowered in representing themselves. They were communicating with those working in public services, engaging in work with marginalised LGBT people, and knew the usefulness of data in interactions with commissioners and services. This had developed in the context of a network of community-based structures working with, and independently of, statutory services. These in turn not only enabled CMIT to happen, they also facilitated the ideal of 'ownership by the community', and secured buy-in from community and voluntary sector as well as public

---

1   Appendix 1 reviews the specific methods of undertaking research.

2   Throughout this book quotes from focus groups, consultations and feedback events are referenced in brackets after the quotes. Interviews are not. All the interviews were undertaken on an individual basis and pertain only to those 'named'. See Appendix 1 for details of the research methods used and the footnotes for details of ethical issues, including pseudonyms and the cleaning up of quotations.

services and policy makers. In this project, these antecedents to the research were not 'given' by academics or established through the work of the research project but rather the project engaged with, developed and drew upon these.

In this section we identify and unpack key elements that shaped and supported this project, which we believe have been essential to its success. In doing this, we decentre the researcher and challenge (participatory) researchers to engage with the context of research projects, recognising that these shape what happens and create potentials for research. We discuss four factors: 1) national and local agendas and the local Count Me In research as a precursor to CMIT; 2) Spectrum, the LGBT community organisation that was the project partner with the university, and its role in engaging and representing 'the community'; 3) aspects of practice at the University of Brighton, where the project team was brokered; and 4) how money for project funding was deployed in this project. Viewing these as key constituents in research that seeks social change, or in UK academic speak, impact, we highlight the collective and contextual nature of participatory research and look beyond the individual researcher/research team.

National and Local Agendas: A Window of Opportunity?

Many of the participants viewed CMIT as a product of the New Labour Era in the UK. Economic and social practice in this era facilitated an engagement with statutory services in ways that would have been unimaginable for LGBT people during the period of Conservative rule that preceded it (see Cooper 2004, Chapter 1). The Local Government Act (2000) required local authorities to develop community strategies to improve quality of life for groups defined as socially excluded, working with other local bodies (see Cooper and Monro 2003). By the time CMIT was conceived in 2005, the new legislative landscape promoting equality of opportunity was becoming established, along with outlawing of discrimination (see Table 1.1). This legislation was understood to be enacted in Brighton (see Chapter 3), creating conditions of possibility for CMIT:

> Martin: We're lucky that it (CMIT) happened at a time of huge cultural change and in the five years of the project, probably the top of the curve in terms of people's willingness, just general services and potential allies out there to help make it work. So we had, just by chance, a Tory gay leader of the council who supported the project. I think despite it being a research project it also had to fit in with other things. The time that we did it in, it looked like everybody was willing to be part of an evidence-based partnership approach – that's what services were saying. (Feedback group, November, 2010)

Shifts in the legislative and cultural contexts created an urgency (and indeed an imperative) for various sectors to develop new ways of working with 'communities' (see Chapter 6 for further critical discussion of this imperative).

Evidence and monitoring were required in order to first establish, and then to meet, core local targets for the 'significant minority' of LGBT people in Brighton (see Chapter 1). This approach could be critiqued as creating target driven, neo-liberal subjectivities and cultures (Fuller and Geddes 2008). However, the need for data, combined with national legislative equalities, offered opportunities for particular forms of research to be undertaken and for relationships to develop. The placing of LGBT work on the 'agenda' was temporary and in retrospect seen as a 'flavour of the month':

> Jo: At the time the Count Me In Too report came out, the LGBT issue on the political agenda was very high up. I don't think that agenda is so high on that agenda anymore. We kind of forget that Count Me In Too is there and we can actually use it to raise those issues. I think that's a shame because it was a really good, or it is a really good, piece of work.[3]

Jo's use of the past tense 'was [a really good piece of work]' juxtaposed with 'is' highlights how research ceases to remain current in relation to its context. Rather than simply being due to the passage of time, this emphasises how particular agendas can be demoted. Correspondingly, we use the past tense to refer to the CMIT research and the data, reflecting that we are writing in a time (2012) when the context has significantly changed. As we will argue below, the temporality and spatial specificity of these possibilities limited the successes of what happened 'after' CMIT. Before this, we continue our engagement with the context of the research, now moving to explore another key feature of the Brighton landscape in the first decade of the twenty-first century: the legacy of Count Me In.

Count Me In: Foundations in Community Activist Research

Count Me In[4] (2000) was a grassroots research project that laid significant groundwork for CMIT. It was initiated by LGBT activists who sought to put LGBT issues on the agenda of statutory services, and to explore 'hunches' about a broader range of LGBT concerns, collaborating with gay men working in newly created public sector roles that could legitimately champion work around the specific needs and experiences of LGBT people. A steering group of activists ran the project and employed researchers who acted as consultants, advising on questions, undertaking the analysis, presenting the results and then revising the

---

3   See below for a discussion of interview data and referencing.

4   There can be little doubt that Count Me In was ground-breaking and hugely valued by many LGBT people. Count Me In can be understood in numerous ways, and is itself a topic for an entire book. Here we can only briefly outline some key aspects of Count Me In to inform an understanding of Count Me In Too, we hope that others will take up the challenge of further documenting this momentous piece of innovative research.

final report (Webb and Wright 2001). This reversed researcher–researched power relation is assumed even in participatory research, where the researcher often works to 'empower disenfranchised people'. In this way, Count Me In included LGBT people on their own terms, and built capacity, expertise and knowledge about research amongst LGBT people:

> Peter: Count Me In One brought a very diverse group of people together to work with each other. Those people learned a lot from that experience and they were very lucky to have the likes of [names]; experienced professionals pulling the strings. Count Me In was unique because it was the first step really. [It] alerted everybody how to play the game. If you're going to affect policy, you've got to have evidence based stuff to run with, and the players understood that.

Count Me In set out to establish a base of information about personal, social and community needs of local LGBT people, using a self-complete questionnaire (1166 valid responses), at a time when there was little 'hard evidence' about the views and needs of LGBT people locally. This research then led to the development of a 'LGBT community strategy' and by 2006, key statutory services had carried out their own independent appraisals of the extent to which the agenda had been addressed (Brighton and Hove PCT 2005, BHCC 2004). This reflected wider political changes that put equality issues and 'minority' concerns on mainstream agendas and saw government-funded services take proactive approaches to addressing inequalities.

Six years after Count Me In, CMIT not only built on the experience in Brighton of 'community ownership' of research, it also worked with a ready group of experienced activist researchers who understood the place and importance of research, operating both within and outside of statutory services. Count Me In had demonstrated failings and absences in local services, which may have contributed to local services' enthusiasm to support and participate in CMIT. Relationships were not easy because of this, but it did mean that dialogue and funding for LGBT research had a historical precedent in the city.

Spectrum: 'The Community'?

Spectrum was formed in 2001 to carry forward the recommendations of the community strategy from the Count Me In research. It became the organisation in the LGBT community sector that worked as a project partner in CMIT. Spectrum developed an identity, position and working practices that enabled it to facilitate the development of structures and relationships within LGBT collectives and between these and public sector services and policy makers. Through these, this organisation promoted the realisation of cooperative engagements with LGBT people and communities. The existence of one organisation with which public bodies could fulfil their duties to consult with LGBT communities afforded different exchanges to those that characterised earlier attempts to address minority agendas – exchanges where groups took turns with policy makers and public service providers, lobbying and then waiting for responses. In the context of CMIT, Spectrum was well placed as the 'community partner' to broker relationships between public bodies, the university and LGBT communities. This was because in the main, CMIT public sector partners had good working relationships with Spectrum, which brought them to the table in a positive and supportive frame. Moreover, the insights of the Spectrum co-ordinator who had worked as an activist in Count Me In, as well as the recognition of the Spectrum 'branding' in the city, were central not only to securing funding, and ensuring buy-in from a broad range of LGBT people and groups, but also to confidence in the research process itself.

The Spectrum coordinator and trustees were clear that Spectrum would facilitate dialogue, communication with the broad grouping of LGBT people, rather than act as representatives of 'the community' (see also Appendix 1). Working in the initial partnership with the University of Brighton, Spectrum guided the nurturing of involvement of a wide range of stakeholders. This enabled access into Brighton's complex and multifarious LGBT communities:

> Nigella: It's not just a way of talking, it's knowing the people, knowing the language. It's like any community, the more developed a community is, the longer it takes to have any cultural understanding of what's going on, and I think it's a measure of how developed the activist community is now that it takes so long to get to hear about it. With time comes complexity.

Spectrum, through its coordinator, sought to embed CMIT in 'the community', co-opting a wide range of individuals already connected with activist LGBT groups and networks. In this way Spectrum promoted 'community ownership' of the process. Whilst the presence of 'community partner organisations' is in itself indispensable, how this role is executed is critical to developing participatory research that empowers beyond a group of 'insiders' in the community organisation (see Appendix 1 for further details on the processes used in CMIT). We critically discuss Spectrum's role in the city more broadly in Chapter 6 but here we turn to another key partner in the research, the university, whose institutional involvement created particular conditions of possibility for this participatory research, beyond the desires of an individual researcher.

University Legitimacy and Facilitative Practice

Had CMIT held a solely 'community' positionality, its capacity for gaining participation and confidence from public bodies may have been diminished, and it could have been less effective at getting the research onto 'the agenda'. The University of Brighton not only brought legitimacy to the project, but also enacted a vital facilitation role. This role was related to this university's desired relationship to 'its communities', where the University of Brighton's corporate plan 2007–2012 stated an objective 'to become recognised, as a leading UK university for the quality and range of our work in economic and social engagement and productive partnerships'. The Community University Partnership Project (CUPP) was established within this university to realise 'third sector' work with the community and voluntary sectors. It included mechanisms designed to enact 'institutional values' of the University of Brighton (Laing and Maddison 2007). Using CUPP as a centralised tool, University of Brighton secured and allocated funding, and commissioned specific pieces of work. CMIT emerged from a Brighton and Sussex Community Knowledge Exchange (one project that was part of CUPP) 'matching event' in 2005 where Kath met with a Spectrum trustee. Spectrum were keen to update and rework Count Me In to inform Spectrum's work in the city and to help improve the 'evidence-based practice' of the public bodies in the city.

University–community power relations manifested in these CUPP initiatives were often presented as uncontroversial and indeed at times benevolent. Overt discussions of power relations between 'university' and 'community' remained subject to censure and reiteration of hegemonic relations:

> Dana: I saw BSCKE (Brighton and Sussex Community Knowledge Exchange) as more on the edge rather than [other aspects of CUPP, where] what academics wanted seemed to be more important than what the community wanted, and that made me really furious. I thought the whole point was we weren't inside and we weren't outside. I thought it was all about deconstructing power. It often felt to me as though the actual reason for the project was to boost the profile of the university and that that's what they were there for. So it wasn't so much about the process as about good content for papers and books and people to build their careers on.

> None of that was explicit, but there was a real sense that you needed to come up with a plan that would be followed and when the community was perceived as becoming 'difficult' [there was a] sense of 'Oh god there they go again.' So [it was] quite a patronising attitude a lot of the time. It really did feel as though it was a university initiative, it often didn't feel like a genuine partnership to me.

> There were some really fabulous academics there who genuinely didn't want to take the lead and wanted to work with community groups. But they would have a lot of anxiety about what impact that would have in terms of what they'd be able to produce, and they often weren't really supported in that.

Dana Cohen, the BSCKE coordinator facilitated the initial meeting and coordinated follow-ups. She sat on a monitoring group for the project on behalf of BSCKE who provided initial funding (see further on in this chapter for more on funding).[5] Dana

---

5   This group had no decision-making powers, but supported the relationship between the university researcher (Kath) and the Spectrum coordinator and lead trustee (Leela, from 2006).

facilitated reflection on processes and progress and enabled important slippages in timescales to ensure the research was done 'right' with proper community buy in and periods of negotiations and not as a 'quick and dirty' consultancy. This illustrates the wider roles that 'institutional actors' can take in the creation of participatory research.

Dana also had the key skill of helping people to discover 'what they already know' (Stoecker 1999: 847) and supported the team to negotiate difficult relationships and tensions, which were part of learning to work together. She acted as a buffer and a supporter of LGBT research in the broader university context. Such actors can effect change, pushing socially progressive agendas (see Cooper 1994) and show the importance of individuals being in the right place at the right time. Despite the increasing emphasis on social exclusion as more than about poverty (see Craig and Taylor 2002, Chapter 6), and the location of the University of Brighton's campus in the gay capital of the UK, LGBT issues had to be 'pushed' internally (for a broader discussion of the exclusions of sexuality from these research agendas see McDermott 2011):

> Dana: [My manager's] perspective I think [was] around poverty being the important thing, not communities of interest. So that was hard, because I was LGBT and I'd worked at Spectrum and then I was pushing [the research]. It felt as though there was a lot that was risky for Count Me In Too. There were two LGBT people on the Board who were inside the university and they were supportive and again it felt really important. I felt if there hadn't been queer representation on the Board it might have made it harder whereas they were able to legitimise it.

> I mean it [the University of Brighton] provided a space, it provided the money, it provided the overall structure, it provided the space and a lot of the resources, including you [Kath], so that Count Me In Too could happen. It couldn't have happened in a community setting because there wouldn't have been the money or the people with time. I mean, my time has been paid for. About two per cent of your time is being paid for.

> It was supportive and interested. As it went on there was a lot of interest in Count Me In Too because it was obviously doing something quite radical and there was a sense of 'Oh God, it can really work'. You can let things get out of hand and it will actually be really interesting and 'something' academically.

Dana's narrative illustrates the importance of university support and its increasing interest as CMIT 'worked' (see below for a critical discussion of this 'success').[6] One key contribution from the university was providing an expert, able to hold

---

6   Although it could be argued that interviewees and other participants should not be taken as 'facts', we see all who participated in CMIT as creating the research. We also

and deploy university resources, along with passion and commitment to the research (see Browne, Bakshi and Law 2009, Browne, Bakshi and Lim 2012). A supportive head of school and head of human geography research backed this project, combined with broader institutional support and recognition. This differed from situations in other institutions/departments, where there can be pressure to get large scale research funding, complete projects and move to the next one, to publish in particular journals and to maintain 'standards'. Such an approach works against the fluidities and possibilities Kath was afforded throughout this project.

Where socially engaged research has to 'prove its place' with funding bodies, as well as local institutions, the bureaucracy of a top-down approach can come into play. During the course of the project, CUPP transitioned from a dedicated, independent funding stream to core funding from the university. As a result CUPP's processes ironically could no longer contest the very basis of vertical working practice (see also Kapoor 2005). Procedures and accounting came from the institution, which usually holds money for research, leaving community partners feeling, at times, impotent in holding the university to account. Entering bureaucratic university systems caused much angst, for example the amounts charged for overheads are, to community partners (and others), staggering. Negotiation of this situation was aided by a supportive head of school who accepted minimal overheads, such that the project was able to offer the same 'overheads' to Spectrum.

For this research, institutional resources, including legitimacy, academic time and key actors, provided a crucial context for undertaking the research, as well as a source of frustration and context in which hegemonic power relations were reiterated. Nonetheless, money and resourcing, the focus of the next section, was a preoccupation throughout the project with CMIT drawing on a number of sources that became available during the lifetime of the project.

Financial Support

In a period where public spending in England increased, evidence based practice was promoted and community involvement was understood to be key in informing the processes of public policy (see also Chapter 6), CMIT was able to lever funding and in-kind support from multiple sources. The project used innovative funding, including funds designed to promote 'community–university partnerships' (provided through Brighton and Sussex Community Knowledge Exchange) and, following this, 'communities of practice' (provided by South East Coastal Communities, a subsequent funding stream for community-university partnerships see http://www.coastalcommunities.org.uk/sussex.html). Spectrum was able to bring a commitment of funding from the local health authority, which initially matched the university's contribution. This partially circumvented the difficulties in securing funding that might diminish participation, and located some decision

---

understand that they can be critical and engage in self-reflection (see Ahmed, 2012) in ways that do not require added academic 'nuance' or second-guessing.

making outside the academy (Pain 2004). Further support for in-depth themed reports following publication of initial findings reports (see Appendix 1 and www. countmeintoo.co.uk) was secured through Spectrum's relationships with public bodies who fulfilled statutory duties through their participation in this research (including the local council and the police). Statutory partners funded the majority of the additional themed analysis[7] and committed public service workers' time to engage in a data analysis process including attendance at analysis group meetings.[8] The conditions of community–university funding allowed important flexibility such that, whilst detailed plans were drawn up when money was 'applied for' by the research team, these plans formed a basis for discussion when action was planned with project stakeholders, and initial plans were not implemented in regimented ways.

Our examination of the antecedents of Count Me In Too could appear rather descriptive. Through this approach we have sought to illustrate the importance of context and address the ways in which research emerges not only from key theoretical or social questions, but also from the alignment of possibilities, here cultural, historical, institutional and financial. Outlining these augments discussions of researcher positionalities and the 'god trick' of creating research from 'nowhere' (Haraway 1991, Rose 1993). In particular, it shows that not only is research created by someone, coming from somewhere, but also that we need to be attentive to the spatial nuances that create conditions of possibilities when, as we often felt about this research, the 'stars align'. As we now move on to explore, such contextualisation allows for nuanced investigation of participatory research success. Moving beyond a focus on set and tangible outcomes that measure the impact of research, we critically delve into the perceived achievements of CMIT.

Did it Work?: Exploring the 'Success' and Legacy of Count Me In Too

Beyond collecting robust, reliable and informative data through reflexive research processes, as a participatory research project CMIT sought to effect social change and, throughout the data collection participants called for more than 'statistics in an in-tray' (Rosa, trans focus group one). 'Positive social change' was not defined, instead it acted as a potential that project participants converged around, it was a 'shared language' that nonetheless refused homogeneity. Using this phrase

---

7   Key in maintaining the participatory nature of the research and working against established hierarchies were negotiations around payment and control. Throughout the process, paying partners did not have decision-making positions or privileged access to data, nor were they able to offer interpretation and responses to data until the project had published its analyses.

8   With the exception of bi and trans people analysis and reports. Anticipating an absence of direct funding for these, the project team set aside project resources to support this.

accommodated different understandings, hopes and values, ostensibly working for the same aim. The definition of 'positive social change' came into question during evaluation: how could this be 'measured'? In this section we draw on the voices of LGBT people to 'evaluate' this research and find that there is no straight forward or coherent response to the question 'did Count Me In Too work?'.

Responses indicated that the conditions that made CMIT possible, in particular flexibility regarding the timescale, support from institutions and having the 'right people', also contributed to its perceived success:

> Dana: It was the most amazing project because it was so organic. What happened with Count Me In Too felt like it was what needed to happen, it was very responsive. It just evolved as a result of whatever meetings were happening with people in the community. The plan basically went out the window. We[9] kept revising it because there was such a commitment to getting as much information as possible from as many people as possible so that it would really be robust and honest and nuanced and true because it was encompassing the complexity of people's experience. That meant it took much longer than it was meant to and a lot more money. I think part of the reason that it worked is because it was so maverick. So it was quite an outsider project I think. There was this sense of 'we're just going to do what needs to be done and we'll find a way of doing it'.
>
> My feeling is that Count Me In Too had a life of its own, that the academic perspective wasn't the container for it. It was more of a support underneath it so that the life could emerge. The research process itself was quite radical [but] Count Me In Too wasn't just about doing the research. There was a reason that that research was going to be done.

For Dana, the 'maverick' and evolving nature of CMIT was its strength; it questioned models of research that seek strict structures and homogeneity. Because the process was seen as successful, this was equated with the success of the project.

Effecting change is not necessarily inevitable as an outcome of participatory research (Pain 2004, Pain and Francis 2003). However changes were attributed to the project. For example, following CMIT:

- There was anecdotal evidence from some local public service providers and policy makers that they felt more confident asking questions that monitored the sexual identities of those who used their services.

---

9 Dana was responsible for overseeing the research as a part of BSCKE, however many who were involved in steering groups, action groups and analysis groups speak of the research in terms of 'we'. This was important to the research team who pushed for the research to be 'community owned'.

- LGBT people were included in key agendas in the city and funding for LGBT some posts (such as an LGBT housing options officer, a GBT domestic violence worker and drug and alcohol intervention workers) evolved from the research findings.
- There was development of understanding of the diversities between LGBT people, particularly in terms of service providers (see Chapter 5).
- The majority of those who completed feedback at project exhibitions about the project findings felt that they were useful and informative (75 per cent), and 35 per cent said that the project had changed the way that they think about LGBT people.

Bi and trans people in particular spoke of feeling listened to and cared for and having their needs and experiences put on agendas that would not have been possible otherwise:

> Eve: Within Count Me In Too there's a massive effort from the start to make sure that all those underrepresented groups were included – especially from the bi perspective – that we were included – and trans. I think that was really the first time, that as a bi group, we felt that we were actually being listened to, that we were part of the community and they did care about us. I think it's made people realise that there are more issues and actually listen to us. Bi and trans working groups come out of that and different groups wanting to actually engage with the bi–trans working group.
>
> I think everyone who wanted to get involved, could get involved. There was something for everyone, if they didn't want to get involved with focus groups, they could do the questionnaire. So it made sure that people could participate as much, or as little, as they wanted, which was perfect, really. Obviously, the analysis as well, it's far more extensive than I thought it would be. Even when we were doing the focus groups to start with about what questions we were going to include, I never actually realised how much work was going to be involved in the analysis and feedback. I never expected to have different meetings for each group, to feedback on those issues, which was good, because it meant you could go into depth, rather than rushing it through one big feedback. So yes, it's gone beyond what I expected, really.

CMIT also challenged the ways in which public sector services had been undertaking 'engagement', as well as up-skilling staff in ways of 'listening'. This was vital in an era where community consultation was required, but often perceived as being disempowering, tokenistic and produced within tight deadlines that failed to effectively involve people:

> Rosie: We want to get community involvement at a much earlier stage of strategic planning. Not just when we've decided what the evidence base looks like and

what we're going to do about it but right at the beginning when we're starting to ask ourselves questions about what needs to be met and even in designing more questions about what the needs are that need to be met. We want to make it sort of modelled for our own internal consultation project. So the officers that have been involved in Count Me In Too have done a great deal of learning in so many ways.

For me personally I've learned a lot more about listening. [I] learned a lot more about how our services appear to people from outside, how the council as a whole appears to people from outside the council. I've learned about thinking outside my own service. I've learned about joint work, about how to relate to people from all different kinds of stakeholder groups, both internal and external. I've learned how to sell difficult messages to much more senior managers who might not have the resources to meet the needs that they imply ... learned how to negotiate those things in a transparent and open way, which has been fantastic for me. Most of all I think we've learned about how to let the community drive the services that the community needs. (Consultation event, April, 2009)

In addition, as Rosie suggested, by encouraging a rethink of the structures of partnership working that many have documented as disempowering (see Chapter 6), research can be deemed 'successful' because it results in change for the people involved (mrs kinpainsby 2008). A number of LGBT people spoke to us about how the research had positively impacted their lives, feelings of self-worth and confidence, in ways that for some tackled isolation:

Davina: When I started out working with the project I had a fair idea about my identity but my idea about my identity has really changed from being part of the project, in terms of owning my experiences and recognising what my experiences are. So it's changed my life, as in it's changed my opinion of my life, but the biggest thing it's changed is there's lots more boxes that I'll tick about things that have happened to me which probably are not that positive. But the box that I probably wouldn't tick now that I would have ticked before was about isolation and I know I ticked a lot of the 'do you feel isolated "yes"' boxes, but by being involved with the project I probably wouldn't now. (Consultation event, April, 2009)

This 'personal empowerment' can also have broader effects. Some described the outcomes of CMIT in terms of how it facilitated engagement between various groupings. Participants told us that they felt empowered to get their needs across and have them heard by those who have marginalised LGBT people, including from within statutory services:

Interviewer: What have been some of the highlights of your involvement with LGBT communities in Brighton?

Ellis: Watching people just cave. Watching people who would previously have put up barriers to understanding LGBT and who just cave in the face of Count Me In Too. Watching people nervously buy into it and go 'oh okay, actually, no definitely'. Watching that change happen. Feeling supported as a bi person in lesbian and gay contexts, or contexts that I would have imagined to be lesbian and gay.

Persia (separate interview[10]): Count Me In Too is a powerful political actor. I have a feeling [about] the work of Count Me In Too which is delight, to tell the truth. For example in the work I'm doing with the PCT – I can go to them and say look here's real stuff to show. Otherwise they'll say 'you don't know, do you?' 'Oh yes we do'. That's really, really powerful.

Although statistics from CMIT were frequently quoted in public sector documents locally and nationally, ensuring that LGBT issues are 'counted' in particular ways, quantitative methods can be exclusionary (see Browne 2008, 2010). Although there was some evidence that CMIT excluded some who felt marginalised by their sexual and gender identity, the project's focus on those on the margins of the gay city meant that many of those 'in the centre' did not engage. Therefore, for some LGBT people CMIT was not 'about them':

Andrew: Count Me In [Too] is a particular cut of general inequality or disadvantage. I have, I don't know, 150 gay friends on Facebook and much of the stuff that's been identified in Count Me In [Too] does not have anything to do with these people.

I think there are problems within the community, but I don't think the community has problems. I think when it's supposed to be about yourself, my community, I don't recognise it at all. I'm wondering whether I'm wandering around in some pseudo hetero denial [laughs], completely missing out on what's going on in the city, but I don't think I am. I mean I go out once a week at least. Seventy-five per cent of my friends must be gay and just none of this rings bells for me.

Just to be really clear, the stuff that Count Me In Too is identifying around the community safety – bullying, accessing services – I accept all of that, those continuing difficulties in places. I haven't got enough angst [laughs]. I know that's the sort of very basics of, you don't carry out research on something that's going well, we don't want to prove something's good necessarily.

---

10   Interviews are not labeled as such, so unless indicated throughout the book when respondents are put side by side their narratives come from separate interviews. See Appendix 1 for full methods details.

From the outset, CMIT sought to capture and report the views of marginalised LGBT people. Whilst we always acknowledged that it was an impossible task to incorporate all who might identified as marginalised, there was a deliberate focus on marginalisation (and 'problem areas') in the reports. This was in part to counter the assertion encountered by CMIT researchers that suggested 'I am gay/I have a gay friend and I am/he is fine', which was followed by the assumption that all 'gay people' in Brighton are fine. The diversity of experience within the collective however includes those who understand themselves as part of, and included in, Brighton (see Chapter 3). As Andrew argues, CMIT reports could then be read as not being applicable to the majority of gay men. Alongside this, there was pressure from some to present Brighton as a 'good place to live', and indeed publications from the project emphasised that the majority of respondents experienced Brighton in this way (see Browne 2007). However, for those seeking to use data to push services to deliver more LGBT specific work, this message is not central, nor was it seen as relevant and it was understood as potentially counterproductive. The extent to which one research project can represent the diversity of LGBT experience is of course limited. There was awareness throughout the project that framing occurred when reporting on data, and that there were conflicting agendas about how the data might be represented. There was acknowledgment that at the same time as failing to attend adequately to the 'positive' aspects of 'gay Brighton' (see Chapters 4 and 5) and people such as Andrew's friends, the research also did not represent marginalised groupings that had yet to be named/categorised, or who had not come onto the 'radar'.

The effect of participation in research (and thus its power) can be associated with the capability to deploy 'persuasion, negotiation, seduction, inducement and manipulation' (Kesby 2007: 2825). This can be used to persuade those with power to make changes that are supported by research findings as well as recruiting others into undertaking the changes identified in findings reports. During the CMIT project, both those demanding change and those charged with its delivery showed strong enthusiasm about undertaking the research. Yet producing and sharing evidence does not necessarily mean that it is used in a way that progresses social change. Peter, for example, spoke of needing to 'kick them very hard', challenging an assumption that if 'we do the research then they give us the money'. He believed that working in partnership was a failure of CMIT, particularly where this form of working did not develop oppositions and enemies (see Chapter 9) or 'tabloid' headings. Similarly:

> Martin: Lots of us came into this believing that evidence was sufficient to move mountains and make a long-term difference. What we didn't get was a process for taking things forward, a new way for service commissioners and providers to engage with LGBT people. In the future I think we just need to think about what practical things we can leave in place, to make change happen more. To maximise the benefit of that dollop of findings [we need] to nurture those relationships between the activists, providers and partners but also [to] reshape

how they engage with LGBT. But what appears to me to have happened is they just go back to business as usual and it just seems really tokenistic to me.

I suppose a failing for me – and I feel it very deeply – is the lack of some process to go forward with that incredible weight of findings. Especially around the huge pockets of disadvantage and isolation and exclusion, which didn't merit a report on their own. But how are things going to go forwards from here? I think a lot of it will get lost which is very sad to me.

Lorraine: We've done the cycle twice now [Count Me In and Count Me In Too], where you think 'At the end of it, we'll know what to do!' and the reality is we're in the same situation. There were some assumptions in Count Me In Too that were different than Count Me In, which [was] if the services are now ready to work with us, if we get them sitting around the table in the further themed analysis we will automatically get onto their action plans. It's worked a bit but I think we are questioning now how sufficiently that worked. (Feedback group, November, 2010)

This discussion pointed to hopes for the CMIT research and eventual disappointment that the research did not achieve all that was hoped for. The production of an evidence base that did not work 'sufficiently' was identified in the absence of processes 'to go forward'. The presumption that partnership working would be maintained beyond the research process was problematic in itself, as persisting unequal power is asserted through the control of resources for continuing work, and demands for 'better'/'new'/'other' data from those who are in control of key resources (see Browne, Bakshi and Lim 2012, Craig and Taylor 2002, Newman and Clarke 2009). When research such as this is 'not working' there is a danger that it saps energy and resources and can 'distract' LGBT voluntary groups from critiquing models of engagement used by the state through policy makers and service providers:

Martin: The distraction for Spectrum – I suppose with Count Me In Too – is that we had to give up the process [of] having regular big open public meetings because we got caught into dissemination events and preparing for that. I've been asking myself what would have happened if Count Me In Too hadn't happened? What would we be doing now? Would we still have had those mechanisms dealing very directly between people in the community and service planners? Because what Count Me In Too allowed them to do in a sense is to disappear into back rooms with a report and then they didn't then have to talk to people or engage with people. We just lost our funding from the PCT (Primary Care

Trust[11]) because they didn't agree with the model that we had around service user engagement. (Feedback Group, November, 2010)

As well as contrasting ideas about models of engagements, there were differences of opinion about optimal ways of gathering evidence and these undermined the possibilities of CMIT effecting change. Some services and commissioners talked of a need for 'randomised control trials' and 'peer reviewed research' as the 'gold standard' of data and thus negated and questioned the findings of CMIT. Cornwall and Jewkes (1995) argue that health professionals are trained in the superiority of their knowledge and spurious notions of objective research, and the 'purity' of science 'which numbs them to the political realities of life in the real world' (1674). In medicalised contexts then, the 'right' researcher with the 'right' form of data was needed to meet the criteria based on medical models. The notion that there is data that 'can' and 'can't' be used in particular ways demonstrates the complexities around perceptions of how research can, is and 'should be' actioned (see also Browne, Bakshi and Lim 2012).

Persia argued that, in spite of some disregarding the CMIT data, its existence was still significant and the hope remained that 'somebody might' use the data. Yet in doing so she assumes some sustainability, that the data might be useful and deployed in the future. However, the national and local context, which was an important factor in making CMIT possible, was altering at the conclusion of the research (in 2010). Thus, the 'sustainable' impacts of CMIT were affected by the specific temporality of the national and local contexts that enabled CMIT to occur:

Martin: [CMIT] wanted to empower people to get involved and it succeeded in that. In terms of leaving a legacy behind of empowered people, I'm not sure it succeeded. But maybe it's just a victim of the times that we're in, rather than a fault with the project.

We anticipated that times had changed and we could leave a living legacy where people could go on having that engagement. I think most of us thought we were living in a golden age and it was bound to stop at some point but that it wouldn't stop so dramatically, with such fallout and recriminations. It's [CMIT's] impact is severely limited by that, because the sustainable legacy of those connections that were created by Count Me In Too are under threat because of external factors. But the legacy that is left is those voices of people saying 'No, it's still not okay. Yes it's fantastic [that] I can have a civil partnership, that I could report this to the police, but I still don't feel safe, I still feel isolated.'

Sophie: The biggest impact is that those facts [are] out there and you can't ever take them away. What you do with them is something else and what happens in

---

11   The Primary Care Trusts were local health authorities that allocated funding and commissioned services in the period under investigation.

> the future is something else. But that paving stone's been put there and you can never ever take that away. (Feedback Group, November, 2010)

Whilst Martin believed the possible legacies of CMIT are 'severely limited … because of external factors', Lorraine read the situation slightly differently. She argued that operating in the 'working with' model can obscure acknowledgement of impacts that the research might have and what it cannot do. In this way the hope for change through research was retained by some and the empowerment some felt through the process of CMIT was partially maintained:

> Lorraine: We invested in a model for Count Me In Too that was about working with service providers and it's different than being in campaigning mode. We didn't go in with campaign statistics, we invited them to come and look at the stat[istic]s with us and it ends up going to the back room and they sort stuff out without telling us. But part of the nature of that interaction is they're not going to come back to you and say 'we did this because Count Me In Too told us.' They're going to take all the credit themselves. In order for the model to work, we have to be gracious enough to let them do that. That makes it pretty impossible to measure impacts in terms of changes that happened because of Count Me In Too. It's the information in the reports that count[s], we can only get as far as we produce that information. We can't say what that information did. Either we play this game or we don't – maybe now we need to go back to campaigning and holding people to account, but at the time we thought that informing services was the way forward and that's what we did. (Feedback Group, November, 2010)

There was clearly complexity in exploring the 'successes' of CMIT. The model of working 'with' (crucial for participatory research) also challenged the extent to which there can be measurable knowledge of the impact that research might have and that which it cannot do. It may be that it is beyond the capabilities of this model of research to measure what it 'can do' or has achieved. Clearly there was an expectation that gathering and sharing information would in itself do something.

For some, the hope was for a better or ideal life that sees the 'end of discrimination'. Others argued for the value of participating in the process of engagement and inclusion. We finish with a narrative that emphasises that the expectations and hopes for social change can create the possibilities for participatory research, but also that the desired societal changes were not realised, or indeed achievable, through CMIT or perhaps any research project:

> Lorraine: Homophobia isn't going away. For me, activism is about surviving as me, surviving through all the shit that happens to me and my friends and everyone I know, and that's what Count Me In Too does. If I was hoping that it was going to take all the shit away, well it's not going to, so I'm not looking for that sort of impact. But if it means that a group of us got together and we

were listened to by services, even if the services don't do anything once they've listened to us, for me that's a pretty major impact to even be heard. But it's not what people are looking for. They're looking for the end of homophobia, the end of homophobic violence, the end of discrimination. (Feedback Group, November, 2010)

Conclusion

In this chapter we have explored how research is formed through its temporal and territorial situation, and examined what happens when 'successes' are explored in their own context. We argue for examinations of possibilities that are grounded in where research takes place, rather than extrapolation of universal models that are presumed to have usefulness across a range of situations. Whilst drawing on some of the elements and practice of participatory research, we contend that conditions of possibility need to be accounted for alongside other aspects, such as the positionalities of researchers that create research. In this way, we further challenge objective research that supposedly comes from nowhere. In contrast we see the possibilities of undertaking this research in participatory ways that cross community–university–statutory sectors as highly specific and thus perhaps irreplicable. We are thus purporting that the transferable understandings of the processes undertaken to produce this research are about the 'frequently idiosyncratic and always performative nature of learning', rather than producing models or 'formulaic distillations... and instrumentalist applications' (Amin and Roberts 2007: 353), which others might use again or deploy elsewhere.

In examining the 'success' of the CMIT research through reflection by those involved in it, we displayed a breadth of perceptions, rather than a unified and coherent portrayal of 'success'. As we have shown, there was purposeful vagueness about what change the project was aiming to achieve, which enabled the recruitment of a wide range of stakeholders with diverse and, at times conflicting, agendas. Feelings of success are, however, only part of the story. Participants referenced what they would like the project to do, namely to eliminate prejudice and discrimination. This emphasises how, for some of those involved, 'success' was not about the project itself but rather about the aspirations for change. The possibilities of what gay Brighton should offer influenced these hopes and were indeed an important driver for change.

We finish this chapter with a different voice, that of Arthur Law, a key activist who gave his time both to Count Me In and Count Me In Too. He made a speech at the launch of the CMIT exhibition of findings during Pride week in 2009. The extracts taken from his speech offer an insight into the successes and limitations of the CMIT research. The imaginings of what Brighton should be and the failings identified by CMIT in attaining this ideal are the focus of the next chapter.

Looking Back to Count Me In

Arthur Law

We started on this journey together because we wanted to make a real difference. Some of us started off thinking we knew what needed to change and why:
 • We wanted the proof of a prosecutor wanting to press charges in court.
Some of us anticipated we would find awkward things that didn't fit:
 • Or find incriminating stuff that people would want to use against us.
Some of us were tired of not being heard or always being misheard:
 • We just wanted to find a voice.
Some of us spoke because we were invited to speak:
 • The injustices we face are so routine that we do not notice them.
Some of us were interested in what was over the garden wall and who our neighbours were:
 • We were wanting to know what was and wasn't going on.

Some of us neither got the invitation, or it was buried in junk mail and spam, couldn't cope, didn't have the party frock we assumed was needed, or just didn't make it.
Some of us, despite our best plans, were still unreachable.
Enough of us made it. Enough of us spoke. Enough of us were counted.
So did we make the difference?
Maybe we are the difference?
Did we create an 'us' that was expanded beyond those who had already staked their flag in the rainbow city?
Did we manage to arrange the room well enough so that everyone could find a welcome and a space?
Did the leaders, planners and chiefs feel welcome at our table and did they leave wanting to come back to talk further or negotiate peace?
You decide.

So what did we find?

 • We found the courage to tell it how it is.
 • We found painful, challenging, heart-breaking and inspiring stories; voices that spoke from the heart, and from the bitter, broken truth of it and from a shared space of belonging and love.
 • We found a yearning and, yes, passion to redefine the shapes and edges of our worlds and for unity.
 • We found different ways in which we fit and don't fit.
 • We found we needed each other and help to reach out to each other in new ways.

- We found the Argus and Daily Mail too distracted to point fingers of judgement or ridicule.
- We found a way to share a dialogue about an 'us' which included 'them'.
- We found tools for breaking rocks, laying tracks and others that look at clouds and out at the horizon.
- We found evidence alone does not change the world.
- We found far more questions than answers.
- Some of us still weren't able to be heard.
- Some of us found ourselves.
- Some of us found each other.
- Some of us found an us.

These were very important findings because, without a sense of who we are and where we want to go, there can be little progress.

Chapter 3
The Promise of a City Paved with 'Gay Gold'

With its laid back, bohemian atmosphere and reputation for cheeky, free thinking, lesbian and gay Brighton has long been considered the UK's gay capital.
– (http://www.visitbrighton.com/site/your-brighton/gay, accessed e22 March 2011)

Robert: No matter where you live in the country, you're aware that Brighton and Hove is the gay 'Mecca' of the UK. So you're expecting the streets to be full of feather-wearing, pink-painted, sequin-encrusted, LGBT people running around. There is an expectation that we're all partying every night; that, as a gay man in Brighton, you are lucky cos you have everything given to you: the social life, the support systems, the acceptance of the wider community. That doesn't exist. It is just a seaside town with a big gay scene.

Introduction

This chapter begins the discussion of place as active in the constitution of ordinary lives, by exploring the power of imagining Brighton as the 'gay capital of the UK', as occurred in the first decade of the twenty-first century. Place[1] can be viewed as creative/active in many senses. We invoke the concept of a 'sense of place' to examine how place can be read as a social relation alongside other forms of social relations (see Hubbard 2006, Soja 2010). Sense of place is a longstanding geographical concept whereby meanings attached to place 'generate spatial arrangements, which then impinge on meaning', such that how place is imagined cannot be separated from how people live their lives (Hemingway 2006: 432, Gibson 1978, McDowell 1999, Relph 1976, Shields 1991, Soja 1996). Sense of place is thus the 'lived, embodied and felt quality of place' that forms practices and identities (Martin 1997: 108). As we make sense of places, we ourselves can be attached to, as well as detached from, them. Thus the imaginings, myths and representations of places, whilst supposedly fluid and ethereal, have real effects in the mundane aspects of daily life (Amin and Thrift 2002, Binnie 2004a, Fincher and Jacobs 1998, Lees 2004, Said 1978, Skeggs et al. 2004).

Our focus on Brighton develops conceptualisations of how perceptions of the sexualities (and genders) of urban places recreate LGBT lives. The chapter firstly

---

1   While cities are not synonymous with place, cities can be explored through sense of place and through explorations of how urban contexts construct our lives and politics.

explores how Brighton was[2] idealised as a place where the streets were supposedly paved with 'gay gold'. We illustrate the diversity of urban imaginaries when LGBT equalities seemingly 'apply everywhere', at least through national legislation (see Richardson and Monro 2012). By exploring what cities do, we answer the call to examine place as active in the reconstitution of social relations (see Hubbard 2006, Soja 2010). We go on to investigate how some LGBT people were not 'in place' in the gay capital, examining hate crime and housing. Our discussion of moving to Brighton without 'housing solutions' offers stark insights into how place recreates exclusions and the material effects of difference. In dealing with the failings of the city supposedly 'paved with gay gold', we paradoxically show the productive effects of such failings. The inability to realise the promises of the imaginings of Brighton as a 'gay Mecca' gave rise to specific rights claims about what the city *should* be doing. In this way, we demonstrate how urban imaginings and failures acted as a catalyst for LGBT politics, in a city that was supposedly 'leading the way' in manifesting LGBT equalities in the early twenty-first century.

'We are Everywhere': Imagining 'Gay Brighton'

> Yvonne: Brighton has a particular characteristic and feel to it. It attracts certain sort[s] of groupings of people, but quite a mixed grouping of people, and that is a real strength in Brighton and Hove. You do get a mix of the different nationalities, people of different sexual orientations. It does have a very creative side to it. It's a very artistic place. So it is certainly unique in the sense of it is right on the south coast, it's a small city for a city and therefore it doesn't have that huge vastness that some other places have. That's a nice part of it really.
>
> I think the critical mass of the people that we have here, in the sense of having a very large LGBT population, does set Brighton aside from other areas. That isn't to say there aren't other places that have that, but I think it's quite different in the sense of the LGBT community does have a voice and is seen as a relatively important contributor in terms of community and economy and so on. I think probably the LGBT voice and influence is stronger here than [a] lot of places.

Yvonne introduces the perceived successes of sexual (*sic*, see Chapter 5) identity politics in Brighton. As a hub for pursuing radical agendas and alternative ways of life (Thomas 2011), Brighton was seen as accepting of gay lives, as part of a broader 'mix', alternativeness and tolerance (see Chapter 1). The 'critical mass' of LGBT people in the 'small city' was seen to give 'voice and influence'. Along

---

2   We use the past tense to discuss the research data because the data referred to relates to the first decade of the twenty-first century. While some of the insights presented here may still be relevant, further work is needed to explore the similarities and differences of the political and economic eras that play a part in constituting gay Brighton.

with this, there was an expectation that voluntary and community sector groups would 'be there' for vulnerable LGBT people, catering for any needs that might arise. This assumption was based on the belief that Brighton had a large LGBT population, which was seen to be a key element in the uniqueness of Brighton. As we will go on to show, the imagined distinction of Brighton compared to other areas and cities had an effect on the everyday experiences of LGBT people and the political possibilities in the city. This section explores how Brighton differs from the urban (quartered) imaginaries that define discussions of gay ghettos (see Binnie 2004) to see the whole city as gay friendly.

Broad imaginings of Brighton worked to shape the experiences of LGBT people in the city, such that place actively operated to reconstitute LGBT lives. In the CMIT research in 2006, 76 per cent (n. 618) of those who answered the question indicated that they found it easy or very easy to live in Brighton and Hove, with only five per cent (n. 41) of respondents rating Brighton and Hove a difficult or very difficult place to live. The perceptions of Brighton as embracing gender and sexual difference were discussed in terms of safety in public spaces and 'belonging in the city'. Thus the 'Brighton bubble' was understood as providing a different world to that outside, and offering protections and the possibilities of enacting LGBT identities that were not available elsewhere. These imaginings of place reformed everyday lives, through understanding that there is the possibility of being 'gay' anywhere in the city, and this was described as a visible aspect of the city:

> Jo: There's a realisation [that] the LGBT population lives right across the city. It's everywhere. It's a part of Brighton and Hove.

A large, dispersed LGBT community was considered by some to be part of the fabric of what made Brighton, and a key aspect of the alternativeness of the city. Thus, the extraordinariness of Brighton was in part defined through the commonplace positioning of LGBT people who were 'everywhere'.

The out LGBT presence across the city contested the gay–straight dichotomy that supposedly defines urban areas (see Browne and Bakshi 2011) and asks for a whole city view of this urbanity (see Brown 2008, introduction). It has long been asserted that LGBT lives are lived mainly in heteronormative contexts (see Valentine 1996), which are often juxtaposed with accepting areas of cities. These have mainly been investigated through the notion of 'gay ghettos'. Brighton contrasted with common readings of cities as having defined 'gay areas' with the rest of the city being read as 'straight' and 'dangerous'.[3] Instead, this city was

---

3 In the first decade of the twenty-first century LGBT people in gay Brighton were visible in neighbourhoods and areas across the city. This was manifest through the presence of LGBT reference points such as businesses displaying rainbow stickers and neighbourhood pubs run by out lesbians who organised LGBT-focused events as well as drag queen shows in supposedly 'dangerous' straight areas. Shops, pubs, community

said to be 'gay friendly' with some taken-for-granted exceptions that included particular streets at night and certain neighbourhoods (which often related to class, see Chapter 4, and Taylor 2009). Casting Brighton as 'gay friendly' effected a reframing of where lesbian and gay identities could be enacted. This included supposedly straight or 'hostile' venues and on city centre streets. Thus, place was active in how sexual lives (often understood only in terms of same-sex enactments, see Chapter 5) were normalised:

> Many venues that wouldn't classify themselves as gay venues are nevertheless gay friendly and often with a mixed clientele of gay, lesbian and straight people where I would feel comfortable. Brighton is a very mixed and accepting place. (Questionnaire 45)

Because Brighton was perceived as being 'mixed and accepting', people used spaces in ways that reflected these imaginings and such uses reiterated this sense of place. This in turn recreated a 'visible presence' that served to perpetuate particular gay/LGBT imaginings of Brighton. However bi and trans identities were not always easily or safely enacted, as we explore in Chapter 5. Nonetheless, it was often assumed that the LGBT collective as a whole benefitted from these imaginings of gay Brighton. Bi and trans people also reported similar imaginings of Brighton and these were often formed prior to living in the city (see also Browne and Lim 2010).

Participants discussed the possibilities of overtly enacting identities beyond social spaces and public streets. Workspaces for example offered some people the possibility of becoming 'normal':

> Andrew: My experience in Brighton has just been fantastic. The level of respect, support, engagement, consideration [and] commitment of key people to LGBT issues is fantastic. Many of these people aren't gay themselves, or LGBT, but [are] absolutely fighting for it. The senior staff [are] just relentlessly driving the agenda and personally supporting me and championing me and mentoring me and helping me to develop – absolutely fantastic.
>
> I don't know whether I'm hugely lucky. It's been such a positive experience. We had another child. It's sort of an unusual situation when you're a gay couple having a baby with another gay couple. They could say 'well you're not entitled to this paternity leave, let's be realistic about the situation' – they're not at all. I mean complete flexibility, allowing me to take [paternity leave] in a way that reflects my involvement with the child, cos statutorily you're supposed to take [this leave] in a block. Coming back to my office full of balloons and a card from

---

centres and regular events enabled some LGBT people across the city to enact their identity and be part of a collective that connected with local neighbourhoods.

the leader of the council, congratulations on the baby [from] the whole team – absolutely amazing I have to say.

It's a little bit sad in a way, the fact that I'm commenting on it, because that's what sort of happens for everyone. It's sad that I feel a necessity to comment on it. I'm bringing it out as an example. Actually [it] is normal social interaction. I haven't noticed any difference in how I've been treated. I mean they're just lovely. Maybe it's just where I'm sitting, in the immediate team that I'm in.

Andrew's experiences and the 'luck' that he was afforded could be viewed as manifestations of white gay male privilege (see Noble 2012). Experiencing 'what happens for everyone', Andrew sees his commentary on LGBT issues as 'sad', indicating that he places importance on further assimilation that would negate the need to even mention the 'amazing' way that he was treated. In discussing the difference of his parenting situation, he says that a 'realistic', or perhaps homonormative, approach mirroring presumptions of heteronormativities might deny him access to paternity leave. Instead, Andrew's narrative outlines implementation of paternity leave arrangements that went beyond presumptions about how 'family life' should be catered for. He describes how gay/queer families contest assumptions about parents and parental roles and yet were catered for, for some. Thus, Andrew sees himself as different but being treated 'the same'. Andrew grounds his experiences in Brighton, which is thus invoked as creating a form of normality, through its extraordinariness in relation to other places.

This extra-ordinariness came with a certain amount of collective power, and 'the LGBT community' was understood to be a powerful economic, social and cultural driver in political arenas such as local government. Momentous occasions for LGBT equalities were embraced in the rhetoric and actions of the local authority. For example, Andrew pointed to the symbolic importance of being the first city to register a civil partnership,[4] opening the registry offices early in order for this to occur:

---

4   There is not scope in this book to address the complex debates regarding civil partnerships and same sex marriage. These often argue as to whether same-sex/gay marriage is assimilationist 'buying into' particular norms and (capitalist) power relations, or whether the presence of two men/two women is inherently disruptive to heterosexual marriage (see for example Warner 1993, Sullivan 1997, Halberstam 2005). Many of these debates are US based. Research on civil partnerships has sought more nuanced engagement with the experiences and practices of how these are enacted and lived (see, for example, Peel and Harding 2008). In this context, it is pertinent that civil partnerships were used and understood as 'progressive' and a sign of the LGBT friendliness of Brighton and Hove City Council, and, as Ellis shows below, those who may be disadvantaged by the legislation were also considered. However, civil partnerships continued to be a contentious issue and it was recognised that many were disadvantaged through their introduction (see Browne, 2011a).

> Andrew: The civil partnership thing [is] for me probably the landmark. It's
> the legitimisation and respect to our relationships. Absolutely amazing.
> Transformational. I think it's really affirming to have that there, really powerful.
> The fact that Brighton was the first one that did, that we really led on that as a
> council. Just fantastic.

> Interviewer: What do you think about legislative change for LGB and T issues?

> Andrew: In Brighton and Hove I would hope it wouldn't be that big a deal. We
> should, and many people are, trying to do those things. It's [the legislation] a
> good stick to sort out those people that aren't. Where there are problems, I know
> the council would thoroughly enjoy doing anything they can and I'm sure the
> police will as well. Elsewhere in the country, it potentially has a massive impact.

Some, like Andrew, felt that the 'stick' of equalities legislation was not needed in
Brighton, outlining a hierarchy where this city that 'led the way' above 'elsewhere
in the country'. Andrew suggests that this is because 'many people are trying
to do those things'. Thus, some understood that the requirements of equalities
legislation of the early twenty-first century were already being put into place in
Brighton, without needing the legislative drivers. This was somewhat supported
by the CMIT quantitative survey data where just over half of LGBT people said
that the city council and other public services were LGBT friendly or very friendly
(53 per cent, n. 403).

For others, the equalities legislation passed in the first decade of the twenty-
first century was used to do more than take care of homonormative people who
might benefit from legislative changes, such as inheritance rights or changes for
those who 'contribute' to the city's image and economy. Equalities legislation
was also seen to create an imperative to cater for those who might be negatively
affected by the legislative changes that benefitted some:

> Ellis: The Civil Partnerships Act has certainly impacted massively on Brighton
> and Hove City Council who are very much pro the Civil Partnerships Act. Also [I]
> think we [at the council] had a fairly good plan B about managing homelessness
> as a result of loss of income, loss of housing benefit income [due to changes in
> assessment, see below]. I was waiting for this really big wave of homelessness
> or people that would need help, but I think probably what's happened is most
> people haven't declared or they've just quietly found themselves more affordable
> places to live or they've moved separately or whatever. I don't know.

> Interviewer: Why do you think [the equality legislation] has impacted Brighton
> and Hove more than other places?

> Ellis: Because there's a really good LGBT community and voluntary sector
> here that isn't going to let statutory services forget their obligations. And there's

LGBT people also employed in the statutory sector and we have vested interests in making sure that our rights as service users are reflected in our duties as a service group.

The urban politics of the 'gay city' were enacted through the provision of mainstream public services (including housing) to vulnerable LGBT people. This was attributed to successful partnership working and the LGBT voluntary and community sector holding statutory services to account, indicating the blurring of state/non-state boundaries and the possibilities of activisms with/within[5] (see Chapter 6). In the case of civil partnerships, there was a strong push from Spectrum to mitigate against the potential losses for those adversely affected by the new legislations (see Browne 2011a, Chapter 6). Ellis spoke of a 'plan B' to deal with the potentially devastating impacts of 'living as if civil partnered' for the purposes of benefit claims and welfare support. This included loss of housing, loss of benefits and other state support from those who were after 2005 considered a 'couple' (see Browne and Davis 2008, Browne 2011a). Contrary to a focus only on those who are 'good gays', such practices indicate a concern for those who need support from the state and an awareness of the negative implications of landmark equalities victories, such as civil partnerships.

The local authority, Brighton and Hove City Council, also provided financial support to the majority of established non-profitmaking LGBT community groups in the city. They received vital financing through grant giving programmes that paid for (a proportion of) core costs. LGBT workers in Brighton and Hove City Council and the financial support of LGBT groups blurred the binaries of gay/straight and state/non-state boundaries (see Chapter 6) in the creation of the gay city:

RC: We have a great support system in our council. It's fantastic to have the power of the Brighton and Hove City Council and the police, the statutory bodies, all those people, working and actively encouraging an event like Pride and thereby helping to support their LGBT community, which is vast, in this particular town. That's great, a really fundamentally positive brilliant thing. Cos these people, quite a lot of them are doing it because they know they have to do it, or they're doing it because that's what their job tells them to do, or they're doing it because they see the benefit of it, not because they're actually part of the community. That's fantastic. They're not LGB or T, but they understand the need to support that community. That's a very powerful position.

Those employed by state institutions supported LGBT groups such as Pride 'because that's what their job tells them to do', moving beyond Ellis's identification of 'vested interests' of LGBT people employed in the statutory sector (see

---

5   Due in part to pressure from Spectrum and other activists, recommendations made for an 'LGBT housing options officer' (Brighton and Hove City Council, 2009) materialised after the Count Me In Too research and the LGBT housing strategy.

also Cooper 1994). Straight representatives of state institutions were seen to 'understand the need to support' LGBT communities and in this way enacted a vision of Brighton that fed from its imagining as an extra-ordinary city where LGBT people *should* be catered for.

Imaginings of what the gay capital *should* be like also drove migratory decisions and, in this way, gay Brighton discursively and materially recreated LGBT lives. Rural to urban migrations have long been the focus of studies of gay and lesbian migratory trajectories, setting up urban utopias that contrast with apparently repressive ruralities (see Weston 1995, Gorman-Murray 2009). Weston (1995) spoke of the migrations to 'the big city' in order to experience the possibilities of the city, as well as to 'escape from' hostile rural environments. Others have critiqued this urban/rural divide and questioned assumptions of rural intolerance and urban coming-outs (see for example Gorman-Murray 2009, Knopp 2004, Phillips et al. 2000). In the CMIT research, the specificities of the place of Brighton even as a relatively small city were seen as central to the 'pull' of the city, rather than the broader draw of 'the urban' and large cities acting in migratory decisions. CMIT respondents noted that people move to Brighton because of the imaginings of the 'pink city', its acceptance, safety, LGBT communities and the support to live LGBT lives that this was perceived to offer:

> Beth: [I] made a specific decision to come to Brighton. I was attracted by the GLBT [gay, lesbian, bi and trans] community. I'd lived in London a lot of my life. I felt the gay community there was not as strong as it has been and I got fed up with having to be at war with the rest of the community. I wanted to come to Brighton, just relax and be amongst people that I felt comfortable with. (Bi focus group)

These narratives were supported by quantitative data. Ninety-three per cent of Count Me In respondents in 2000 (Webb and Wright 2001) and 92 per cent of Project Zorro respondents in 1998 (men only) (Scott 1998) said they had moved to Brighton and Hove/Brighton. Both studies asked respondents their reasons for living in the city and found that the most common reasons given were LGBT/gay scene/community and gay/LGBT friends/partners.

Imaginings of 'gay Brighton' were active in the reconstitution of everyday spaces, migratory trajectories and workplace practices for some LGBT people. However the utopic visions of gay Brighton as 'sorted', open and tolerant were contested and we now look at this. We show that place created multiple experiences of the gay capital, recreating diverse LGBT lives, migrations and experiences.

Streets that are 'Paved with Gay Gold'?: Living in and Moving to Gay Brighton

Munt (1995: 107) contends that the paradox of hope for urban lives is that 'more pleasure is taken in journeying towards it' than arriving. For many, migration to Brighton was driven by idealistic imaginings of the city that culminated in disappointment (see Shields 1991). Not all LGBT people were ordinary here, and the majority of CMIT respondents (73 per cent, n. 596) said that they had experienced some form of abuse over the past five years (2001–2006) because of their sexual or gender identities. Regular experiences of abuse are often ignored or downplayed, enabling not only the reiteration of Brighton as 'better' than elsewhere (see Browne, Bakshi and Lim 2011) but also of the associated freedom of sexual (*sic*, see Browne and Lim 2010, Chapter 5) enactments that rely on this place-based imagining. Accounts of experiences of abuse illustrated the perceptions of the city and its failure to achieve the promised ideals:

> Matt: [I have been] harassed by a bunch of people in our street that for some reason don't like us and I think it's because we're gay. I can't believe that because it's Brighton for god's sake. It does happen, people just don't like you. But we've had so much trouble over the years, we have been called paedophiles, we've had our front door kicked in. (Hate crime focus group)

Matt expressed disbelief because Brighton *should be* doing something to ensure that gay men, such as him, do not have 'trouble'. Yasmin sees Brighton as reconstituting everyday lives in a way that questions the positive assertions of the city as protecting all who are 'gay'. She pointed to the 'soup of oppression' that can be realised only when a tangible or nameable factor is present (see Browne, Bakshi and Lim 2011):

> Yasmin: It is a very subtle thing because discrimination is a word that sounds like something very active that somebody does to you. Oppression, which I prefer to speak about, is like a soup that you are sitting in. It is in your eyes and your nose. It is everywhere. So it isn't something that you can say 'well that person did that thing to me'. You can identify those events but they are like the carrots in the soup, they are like the big bits that you can grab hold of and say 'well I was walking along the Level and somebody hit me over the head without provocation and then proceeded to racially abuse me'. That is a hunk of carrot or leek or whatever that's in the soup, but the rest of the soup is there all the time. That is actually what life is like when you are a woman, you are a lesbian, you are Asian, you are Muslim, you are all of those things that I am. I think in Brighton there is a kind of naivety, 'oh we are all very nice in Brighton and therefore we don't discriminate against anybody'. I think that's a problem. There is a kind of naïve collusion with institutional and other forms of what is soupy oppression, which people don't really recognise. (BME focus group one)

Yasmin challenged the idea that hate crime was the only form of oppression in operation and argued instead for an understanding of the insidiousness of 'oppression', where (some) LGBT people did not become ordinary. In her discussion of intersectionalities, including sexuality, gender, ethnicity and religion, and experiences of the gay city, the multiple and intersecting ways in which some LGBT people failed to attain ordinariness are illustrated (see Browne et al. 2010, Taylor et al. 2011). For Yasmin, assumptions about Brighton's tolerance, lack of discrimination and 'niceness' was a form of 'naïve collusion', one that failed to recognise the ways in which the city perpetuates 'soupy oppression'. The erasure and denial of 'soupy oppressions', including gender and sexual difference but also moving beyond them to encompass aspects such as race and religion, emphasises the heterogeneity of LGBT people, in (white dominated) gay Brighton. More than this, assumptions regarding Brighton played a part in reconstituting these social relations. What Brighton 'is' came further into question when examining the experiences of those who migrated to the city 'paved with gay gold':

> Peter: People are drawn to Brighton because of the perception that there is a gay community here. They think the streets are paved with gay gold and they quickly learn that that's not the case.

> RC [separate interview]: People always seem so shocked that homophobia or biphobia or whatever, still exists in this town. Of course it still exists. It's a tragedy that it exists anyway, but everyone seems so shocked it still exists in Brighton and Hove, cos [it] is such a tolerant and understanding LGBT Mecca and all this stuff. Well it isn't. At the end of the day the bigots are everywhere.

In their individual interviews both Peter and RC pointed to the dissonance between 'golden' urban utopia and lived experiences of abuse that were everywhere. Contesting the extra-ordinariness of Brighton, these narratives placed bigots everywhere. This challenged hierarchies that understood Brighton as better than, and different to, other places. Peter's contestation of the affluence and 'streets paved with gay gold' points to the ways in which the illusion of the pink pound was powerful in migration decisions but lacking in reality. This parallels the illusions and realities of local government 'acceptances' being also only available for some (see Cooper and Monro 2003).

Although LGBT equalities were supposedly being enacted throughout England in the early twenty-first century, there were uneven contours in the enactment of legislation (see Monro 2010). Throughout the country there were sexual and gender intolerances, and there were geographical differences between expectations of LGBT friendliness from local government and public services in different places. As we have seen, Brighton was a place where there were expectations and representations of acceptances and freedoms. Yet, there was a dissonance between tourist and election rhetoric (and the ideal gay evoked through this) and the realities of public/mainstream service delivery, which could cater for LGBT people who needed support. These realities contrasted starkly with the assumptions that Brighton could and would cater for vulnerable LGBT people:

> Jack: I think there are a lot of great words about LGBT in Brighton. They do include us in pamphlets and what have you, but I think actions speak louder than words. When it comes down to the actions, they are not very gay friendly at all. I get the feeling that we are being used simply because it is an attraction here. If you ask anything of them other than to bring the pink pound here and have a good time, don't!

> Dan: I think what it is they want is our money. But they don't want us here.

> Jack: They give us all these fancy speeches and fancy words but I find them full of hot air when it comes right down to it – homophobia, big issues that affect the gay community; they really don't want to know.

> Dan: The message is that you are welcome to come to Brighton but you must have money. You must have somewhere to live and you must have a job. Don't come here unemployed. Don't come here disabled. Do not come here if you have got mental health issues. That is the feeling from the members of staff of Brighton and Hove Council over the last 15 years, the attitude that I have received from all of them. They don't want us here, they don't care if you've lived here all your life, go and find somewhere else, you are disabled and they don't want you because they can't cope with you in Brighton and Hove. (Disabled focus group)

This discussion not only produces an imaginary of Brighton but also describes place enacted through 'them' (Brighton and Hove City council) and thus creates LGBT lives (us). For Dan and Jack the city was embodied through the 'them' that wants 'our money', but not 'us' that asks for 'things'. Expectations built up through 'fancy speeches and fancy words' are met with realities of support systems that 'can't cope with you' even if 'you've lived all your life here'. The latter is important as there is an imperative for local authorities to cater for those who have local connections to the city. However, even those who had particular claims on the city, both in terms of local connection and LGBT identities, found that their needs were overlooked despite the imaginings of a city 'paved with gay gold'.

Participants working in public sector services pointed to the stark issues for those who migrated to the city without a local connection or any financial or housing security. This questions the ideals of choice that can be associated with urban migrations and points to the privileges of safe and positive mobilities that may be assumed for all who move to gay utopias (see also Gorman-Murray 2009). Housing vulnerabilities were shown to result in, and from, following the path to Brighton (Browne, Cull and Hubbard 2010, Cull, Platzer and Balloch 2006). This pertained to broader housing issues in Brighton, which has a relatively large homeless population. In 2003, it was anticipated that the national legal demand for a local connection in order to provide housing support would do little for migratory motivations and people would still migrate to Brighton (May 2003). This research to an extent showed that this requirement did not deter LGBT people from running to the gay city.

Allocation of responsibility for provision of welfare resources for homelessness and other support in England depended on a geographically understood notion of community, where individuals were part of 'the community' that they were born in, worked in, had family in and lived in. In contrast, LGBT migrations can be based on understandings of home and community in ways that contest simplistic links that map geographies of birthplace and family homes to belonging, connection and community. Instead these 'culturally belongings' related to gender and sexual identities (Weston 1995):

> Lee: I got an email the other day from someone who was experiencing hate crime in [names county], trans-phobic hate crime. Basically, they had told this trans woman 'maybe [name of county] isn't the place for you, I think you should move to Brighton'. I told her off for telling her that. Actually you should be making her safe where she's chosen to live.

> That presents a problem for us in Brighton. I deal with a lot of people who turn up at Brighton station without any planned move; because they culturally want to belong here, they already feel like they've got a foothold here because they've gone out on the scene and they've liked it, or because they've heard stories about how wonderful Brighton is, or it could be they're fleeing violence, [or] where they are doesn't feel particularly LGBT friendly. They don't have any cash. They don't have a job down here. They end up sofa surfing. They get into all sorts of scrapes, having sex when they're not actually too comfortable with that, all the kind of things in relation to that. Trying to get their housing sorted out can be really, really tricky.

> There's a slight disconnect between the perception that other areas have of us and the experience people have when they turn up here without a planned move.

Brighton in Lee's narrative is a place that some LGBT people (and indeed service providers in other areas) feel is 'home' even before they arrive. The gay capital implies connection to this locality for a range of LGBT people and that is understood as a local connection. These cultural imaginings are obstructed by housing services who do not view this cultural belonging as a legally defined local connection when dealing with individual cases. Yet in Lee's account, there is evidence that individual workers within services not only assumed but also promulgated a connection to Brighton on the basis of trans identities that were seen to be out of place in other areas. Thus, Brighton is both imagined as a site of sexual and gender freedoms, and different from other areas. However, the ideals that these imaginings were built on were not attained.

During the period of this research, house prices and rental prices in Brighton were rising steadily year on year. This is set in a context of low wages in the city and a high proportion of accommodation in the private rented sector (kept buoyant by the universities in the city), with relatively low provision of social housing (May 2003). This problem was exacerbated for younger unemployed people, whose Single Room Rent Allowance did not match the rental prices in the city. This created what May terms a 'crisis of homelessness in the city' (2003: 31). Building on this, Jackie's account eloquently points to the multiple and intersecting issues of sexual and gender identities, violence, class, housing and age that are 'uniquely dangerous' in Brighton. In this way she emphasises the importance of place as a social factor that creates the material conditions of everyday life:

Jackie: LGBT people who migrate to Brighton and Hove and don't have a housing solution are running enormous risks because they didn't move here in a managed way. Because statutory services are not set up to help them [and] community sector services are also not set up to help them. The majority of any sector don't understand the way that identity and needs work together. So you have got a uniquely dangerous situation because of the housing market here, the expense of the private rented sector [and] the way that housing benefit doesn't give young people any equal stake in benefits.

[And] Because of a whole bunch of other kind of intangibles as well. Because of the way that youth is valued on the scene. Because of that whole weird discourse around the sexuality of young men. Because of the way that young men in Brighton and Hove seem to think that they can run sexual health risks. Because of the way that young trans people are fetishised. There's a whole nest of issues focused around a person's inability to find and keep secure appropriate housing where they experience safety, somewhere to regroup and deal with those issues. If housing solutions were provided then those risks could be much better managed.

There's also this other stuff around what happened to those people before they came here. What was so painful about their family life or their home life or their school life that they fled without anywhere to come to? What drives people to just sleep in the park after Pride and never go home? What drove the person that propositioned me at like three o'clock in the morning after Pride last year, myself and my partner, someone who's like maybe 14-15 years old? We were kind of watching him because I was like, 'there you go, duckling'. He sort of saw us looking and came over and asked us whether we wanted to have sex with him and how much we would pay him. What's he running from? What's he frightened of? Why can't he go home? That stuff is the harder stuff for me.

Jackie's narrative illustrates the multiple ways in which place interacts with individual life trajectories and experiences. The place imaginings of Brighton as 'paved with gay gold' contradicted with the materialities offered by the city to LGBT people who migrate without 'housing solutions', especially where accommodation is expensive and housing support is based on local connections that do not account for LGBT imaginaries of 'home'. Whilst others have recognised the ways in which 'homes' can be dangerous places where sexual identities are hidden or indeed punished (Valentine and Johnson 1995), Gorman–Murray (2008) has pointed to how lesbians and gay men can create and sustain homes where they feel safe and able to express their identities. The privilege of taken for granted access to such securities often goes unrecognised. Placing key importance on secure appropriate housing, Jackie understands the potential effects of an absence of a 'home' to 'regroup and deal with those issues'. This was not attainable for all LGBT migrants to the gay capital (see also Cull, Platzer and Balloch 2006). Thus the role played by the city in creating LGBT lives varied in terms of migratory trajectories, which in turn informed who became ordinary in Brighton.

For some, the issue was not with a (heteronormative) 'them', but instead with differences within those who were supposed to be part of 'us'. It was apparent through much of the research that power and belonging were used by some gay men and lesbians in ways that made other LGBT people feel 'out of place'. Bi and trans people in this research spoke of feeling marginalised by lesbians and gay men in supposedly 'LGBT' spaces and in mainstream heteronormative spaces, institutions and contexts (see Chapter 5). Some women spoke of concerns about the silencing of lesbians within LGBT concerns (where bi and trans issues are on the agenda of mainstream policies and services) because of the conflation of lesbian/gay privilege. The power of gay men was mentioned not only in terms of having access to resources such as money and stable housing, but also in relation to how certain gay men influenced and controlled the form and substance of gay Brighton. Certain men appointed themselves as distinguishing good from bad in LGBT organisations and used their position to expose and police those who were believed to be ineffective. They were seen to hold, wield and deploy power in the context of 'gay Brighton' and within 'the LGBT community', and thus recreated the form of these groupings and representations (see Chapter 8).

Despite these differences and power relations enacted between LGBT people, which we interrogate more closely in Chapter 8, claims for trailblazing on LGBT equalities were predicated on an understanding of something that could be labelled 'LGBT communities' and be represented as such with attendant 'needs'. Not all agreed that such connections existed, and Louise argued there is no 'such a thing as a lesbian, gay, bisexual, transgendered community, [rather] there are people who are lesbian, gay, bisexual or trans living within the city'. Thus, who represents 'us' was a fraught and power laden process that was often brought into question:

> Mabel: White men still exert their power and control in the majority of circumstances we find ourselves in, some of the 'activists' are the very worst! The most powerful figures in the LGBT community are men? On the commercial scene, men? I think to speak of a 'community' is to reify an illusion.

I find it frustrating when I hear 'activists' say to, for example, commissioners 'the LGBT community think this or that'. Or 'bi people say', etc. One of the first items in the working agreement we begin our training delivery at [name of organisation] with is that we will use 'I statements' and not generalised statements like 'gays think'. How do you know that, what is your evidence? It is not good enough to just point to CMIT. The data is too complex and the sample source so specific, it should be used with care.[6]

Mabel calls for data to be used with care to ensure that power relations were not reiterated and the illusion of community reified. In the first decade of the twenty-first century in Brighton it was politically productive to argue for LGBT people using data that forced action from local government and others (see Chapter 6), but this may also elide and neglect the heterogeneity of LGBT 'communities'. The misogyny and male power that we discuss further in Chapter 8 can leave some women feeling 'hung, drawn and quartered by their own community' (Melissa). Here it is important to note that the failure to attain the ideals promised by gay Brighton was not only related to the reiteration of heteronormativities, but also related to the hierarchies and exclusions from within LGBT communities (see also Chapter 4).

Thus, gay Brighton was created through complex and nuanced sets of hetero/homonormativities that hierarchised LGBT people, which we will further unpack throughout this book. The effects of this for many of our focus group participants were devastating. They spoke of their disappointment of living in Brighton and not experiencing it as accepting and open, somewhere you were part of and commonplace. They were distressed at how they were treated in Brighton, both in terms of abuse from 'them' and when discussing power relations between 'us'. Many followed the dream of the gay capital paved with gay gold and even here they did not feel accepted and part of the 'community'. This experience can be individualised and internalised as what is wrong with me or you. Rather than seeing the broader social normalisations, such perceptions created emotional turmoil because they did not find the gay capital and had no hope of somewhere better. After all, Brighton was 'leading the way' and as a coastal city was metaphorically understood as the 'end point'. Seeing sense of place as not only forming positive experiences and attachments, negative emotions were also important in considering how cities such as gay Brighton can recreate lives through social relations and regulatory norms.

---

6   Mabel conducted some of her interview by email. This is an extract from an email response.

If it Can't Happen Here, Where Can it?: Politics and LGBT Communities in the City Paved with Gay Gold

Discussions of homophobia, biphobia and transphobia in the city challenged the hierarchisation of Brighton as better than other places and suggested instead that Brighton was 'like everywhere else'. Experiencing the city in this way served to reiterate expectations of what Brighton *should* be and the lives that the city *should* offer to LGBT people. These expectations had powerful political impulses in the UK"s New Labour Era, defined by its expansion of social exclusions to include sexual and gender difference (see Chapter 6). In Brighton practices and policies that sought to work towards social inclusion for LGBT people were in part created through discord in imaginings of Brighton and the hetero/homonormativities experienced in the gay capital. Monro (2007) argues that examples of good practice can be used to 'shame' other local authorities elsewhere.[7] We extend this idea to contend that place-based imaginings that saw Brighton as leading the way were more important than comparisons with other places in the drive to do 'better'. This had multiple effects in urban practices and politics. We focus on the effects of dissonance between ideal of a city paved with gay gold and the multiple and painful realities for those who did not become ordinary in Brighton. This was crucial to how LGBT politics in gay Brighton were manifest.

Despite recognising the failure of gay Brighton to live up to its ideal, participants continued to see Brighton as different to, and better than, other places. This could be read as a failure to see failure as negating possibilities. The hope that the city could provide better lives for LGBT people was key to this:

> Yvonne: I don't think Brighton and Hove are any different to any other city in the sense that there is always a proportion of the population that does not accept LGBT people. It is still a case that people will not walk down the road openly holding hands, particularly gay men. People will suffer from abuse, homophobic abuse. There are instances of homophobia and attacks. Whilst I believe Brighton and Hove is significantly more tolerant than many places, there is still, unfortunately, some way to go to be entirely there.

Yvonne and many others not only understood Brighton as better than elsewhere in spite of the abuse they identified, but they also saw the city as creating possibilities of achieving the ideal portrayed. This feeds into an imagining of Brighton as better placed to deal with discrimination and providing a 'better' experience for LGBT people, even when heteronormativities are apparent. Various factors are used to underpin specific rights claims for LGBT people, including a large LGBT population (that no one can prove does or doesn't exist, see Chapter 1), visible sights/sites such as same sex couples holding hands in the streets and on trains

---

7   See Munt 2007, Probyn 2005, Sedgwick 2003 for a discussion of the productive possibilities of shame.

and the presence of 'the scene' (see Chapter 4), as well as LGBT community and voluntary groups. Conceptualising 'the LGBT community' as having influence can lead to high expectations:

> Audience member: I think fundamentally most of us have a kind of high expectation of what can happen because we're living in Brighton and Hove. I think most of us believe that if it can't happen in Brighton and Hove, where can it? (Consultation event, April 2009)

Halberstam (2011) seeks a reconsideration of failure in light of the normativities of success. Here, failure was productive for activisms and demands for accounting for LGBT in policies and service provision. As Cooper (2006: 929) notes the processes of lesbian and gay work were 'invariably optimistic, not simply despite but also through ... [the] inevitable failure to achieve ... [their] goals; so that "coming up short" became the opportunity to do more and better'. The imagining of Brighton as not only LGBT friendly, but also alternative, cosmopolitan, tolerant and progressive, was used to make the case that the city *should* act to improve the lives of LGBT people when it failed to live up to these ideals. Thus, failed expectations created an imperative 'to do more and better' because of the 'high expectations of what can happen because we're living in Brighton and Hove'.

Yet, as we have started to consider and will discuss further in later chapters, the 'we' that supposedly constitutes 'the LGBT community' is diverse. This led to fragmentation where public sector services, voluntary and community groups and individuals in Brighton grappled with diversity amongst LGBT people, trying to work through systems and processes that asked for consultation with and inclusion of a notionally homogenous community (see also Richardson and Monro 2012, Tett 2002). In addition, there can be little doubt that 'gay people working together' also encompassed tensions, conflicts, hierarchies, internal fighting, bullying and baggage. These power relations were apparent throughout 'the gay community' in gay Brighton (see Chapter 8).

In spite of internal divisions and hierarchies, others spoke of 'gay communities' in Brighton working together. For many of our interviewees, the power of the LGBT collective – rather than gay power – was formed by strength of numbers that was possible because of Brighton's presumed LGBT population size. For example, being heard as part of a broader LGBT collective was key to some trans activism in the city (see Browne and Lim 2010). It was apparent throughout this era that LGBT people sought, and worked to create, a 'better Brighton', and in this way Brighton was not only created, it was also constitutive of hope and political actions. This did not involve all LGBT people yet such actions and hopes were created through imagining Brighton as a place where there is the possibility of making things happen.

We finish this section with one example of the possibilities of exploring an 'us' across differences. This example points to the potential of community bonds. Although the hope and promise of place will always exceed what is or can be realised (Anderson 2006; Lim 2007), hope remains a powerful driver of activism, because failure allows for possibilities. Failure assumes possibilities, particularly the potentials to 'not fail'. Moments where Brighton lived up to its idealistic imaginings offered glimpses of not failing, but couldn't be considered 'successful' in heteronormative terms (see Halberstam 2011). At the CMIT consultation event, 1 April 2009, some participants argued that the speakers' corner, where Brighton-based LGBT people spoke about their experiences of the issues raised by CMIT, created community for them. As Ahmed (2012: 22) argues 'when diversity becomes a conversation, a space is opened up'. In this context, the hope of what Brighton could achieve in creating LGBT lives was momentarily realised, through the possibility of working together if people 'are up for being part of that'. For at least one audience member an 'us' could be created by believing and hoping that it exists:

> Audience member: When we had that discussion around trans [domestic violence] which could potentially have been quite explosive, [it] was actually really healing. Cos I think you've got people in the room who are exploring what an 'us' means, seriously addressing that. If people are up for being part of that, then it's possible. (Consultation event, April 2009)

Conclusion

In this chapter, we have demonstrated how sense of place creates LGBT lives and political impetus. We explored how representations, meanings and imaginings of gay Brighton shaped LGBT lives through urban practices and policies. The imagining of Brighton as a city paved with gay gold was powerful in constructing it as different and extraordinary. Imaginings and representations of Brighton as 'better' created LGBT enactments in the city, informed migrations to the city and reconstituted LGBT politics.

It was not only in positive representations that Brighton, as a place, was seen to be active. Taking place seriously augments discussions that have documented the uneven ways in which LGBT people experience equalities and civil rights (see for example, Hines 2007, Monro 2003, 2005, Richardson and Monro 2012, Seidman 2002, Taylor 2007, 2009, Valentine 2007). Exclusions, marginalisations and otherings were apparent in the gay capital and need to be attended to. LGBT people could experience abuse 'everywhere'. However, rather than solely protecting the 'good/homonormative gays', there was a desire to work with and use legislation to address issues such as LGBT homelessness in Brighton. Whilst such contentions might seem descriptive of one city, detailing this city illustrates that alongside social differences, the specific-ness of place plays a part in creating everyday lives. Taking Brighton on its own terms (see Robinson 2005) offers insights that can be missed in striving to make theories and activisms 'internationally relevant', yet fail to explore the complexities of cultural location.

Following Sedgwick, an examination of the more-than-critical can reveal how LGBT people can 'succeed in extracting substance from the object of culture – even a culture whose avowed desire has often been not to sustain them' (2003: 151). From despair and failure, possibilities and imperatives to do better, beyond heteronormative orders can emerge, there is potential in failure, as well as failure being a political project that disrupts heteronormative success trajectories (see also Cooper 2006; Halberstam 2011). In this research, politics, practices and rights claims were enabled through the idealisations of Brighton and the failure to achieve them. Expectations of what Brighton *should* be meant that the city itself played a role in constructing activisms (see Chapters 6 to 8). These did not necessarily seek to disrupt political orders, acting in resistant ways, nor were they untainted by homonormativities, hierarchies, fragmentation and power relations between LGBT people. In the two chapters that follow, through exploring scene spaces and then trans and bi lives, we develop this understanding of place as a social relation, examining the ways in which Brighton created inequitable hierarchies that spatially reconstituted LGBT lives.

Chapter 4

The Gay Scene: Having it All?

Brighton's community-focused, massive annual Red Party fundraising event is at last here! ... The eagerly awaited explosion of sexy fun and red glamour ... epitomises the city's premier party of the year where the whole community comes together as one to mark world AIDS day in a uniquely Brighton way ... fundraising madness [in] luxurious space to suit everyone!

Packed with top DJs, sexy dancers, outrageous drag diva hostesses and performance artists, it's that one time of year when you can show your support for all the fantastic work done by the city's charities and worthy causes just by coming along and having a good time!

Don't forget to wear head-to-toe RED regalia for this special occasion or show your support by at least wearing your red ribbon with pride!

Or simply lounge around the most luxurious seating areas and numerous swishy bars.

– (Aeon Events, Wild Fruit Red Party 2010)

Persia: I don't think in terms of going to LGBT places. I mean I just live a life. Most bars and clubs and places in town are gay men places. That's how I see them. One or two for the women, but not much. I mean, I don't feel particularly comfortable. The main issue down there in [name of lesbian club] was [that I was] at least 30 years older than everybody else in there. Not for me. And what else is there? So for me there is no scene.

Introduction

Whilst much UK research has explored scenes in London (Soho), Manchester (Canal Street) and to an extent Newcastle and Birmingham, Bassi (2006: 213) contends that each individual commercial gay scene 'has stories to unravel and a task of demystification to attend to'. It is necessary to attend to Brighton's gay scene in order to understand the symbolic imaginings of gay Brighton, and insights more broadly into these key (imagined) spaces of LGBT lives. For some, gay Brighton was synonymous with the gay scene and these spaces were key to how the city was re-imagined and practiced. In turn scene spaces were formed through being in Brighton. Discussing the whole city (Brown 2008) does not negate or ignore scene spaces. Rather it attends to the scene not as the city, but as a part of it. Brighton's gay commercial scene was made viable and reconstructed by drawing on imaginings of the whole city as fun, exciting, risqué, alternative and vibrant (see Chapter 1).

Gay scenes can be understood in various ways (see, for example, Valentine and Skelton 2003, Levine 1979). Here we define gay scenes as commercial venues and events that cater for a 'gay' clientele. Gay ghettoes, scenes and villages are not nationally or internationally homogenous[1] and even within an individual city such as Brighton there was little uniformity to 'the scene'. Gay scenes are therefore better understood as heterogeneous assemblages of emotions, meanings, cultures and materialities that are produced through an illusion of homogeneity. Beginning from the premise that the scene is a heterogeneous assemblage does not negate the importance of the illusion of its homogeneity and, as we will show, its supposed coherence plays a central and, at times, contradictory role in the construction of LGBT lives.

In this chapter we firstly conceptualise how belonging in these spaces enabled some to partly realise some of the ordinariness promised by the gay capital of the UK. As Brighton's gay scene ostensibly created separate spaces from mainstream cultures, it could be read as challenging the ordinariness of LGBT lives which are supposedly part of the fabric of this city. The gay scene does not necessarily follow heteronormativities, but instead questions them, creating alternative forms of belonging and inclusion. However, as we will see, these were not accessible to all. We demonstrate this by augmenting previous discussions of marginalisations from scene spaces in relation to two social identities: the multiple marginalisations experienced in relation to lesbians and the gendering of sexuality, and the intersections of class and LGBT lives. These are both relatively understudied given their salience in creating contemporary scene spaces. The quantitative data in CMIT unexpectedly identified a practice that offers new insights into the drinking spaces that are key to the scene in Brighton (and other gay scenes): specifically, that *not* drinking can act as a form of exclusion. The chapter finally moves to examine debates regarding the commercialisation of sexualities through scene spaces, by considering relationships between 'community' and 'the scene' in Brighton. In this way, we explore ordinary lives through the ways in which 'fitting in' can recreate a form of ordinariness, without neglecting the normalisation of bodies, identities, practices and commercialisms in and through scene spaces.

Gay Scene Matters: Positive Belongings

> Lee: Those bars [on the scene] are 'gay land'. Those businesses determine what LGBT culture is for those people. I think a lot of people feel they don't really have ownership of that space and so it's something determined by other people.

---

1   Indeed the focus on the scene to define gay/LGBT spatialities can rely on specific metropolitan global north constructs of LGBT socialising and community (see Lewis 2013, Visser 2003, 2008, 2013).

They would love to be part of [it] but aren't really. I suppose business is very important, economically, very important in terms of LGBT culture, but it doesn't stop me feeling quite ambivalent towards them.

In academic writing, the LGBT scene has been influential in understanding gay ghettos, lesbian and gay identity formation, accessing 'safe' spaces to be yourself, and developing connections, links and communities. It holds opportunities to perform alternative gender and sexual identities and sets up resistance to police and other forms of violence, discrimination and prejudice (Levine 1979, Kennedy and Davis 1993, Valentine and Skelton 2003). Positioning the scene as more than normal bars, pubs, clubs, shops and nights out, draws on and affords cultural, social and political importance to these spaces (Holt and Griffin 2003: 419). Commercially run venues have been, and continue to be, the main providers of 'public' (these are more often semi-private and profit orientated) 'community' spaces for gay people, and are a key locus for coming to and creating socially intelligible gay identities and participation in gay and lesbian (at times LGBT) communities. Although the place and importance of scene spaces in the global North are increasingly being questioned as equalities are equated with territorial, as well as legislative, inclusions, such that lesbian and gay spaces can be understood as both homonormative in a queer era and unnecessary where other (formerly dangerous heterosexual) spaces can be used openly (see Nash 2013, Lewis 2013). These have geographical specificities and in Brighton Lee and many others suggested that gay businesses and scene spaces acted as a shop window, influencing and creating LGBT/gay cultures through their visibility as venues in urbanscapes and advertising and through invocation of the collective of gay businesses as 'community'. The scene was clearly important for some LGBT people and cultures as it created a space that reamained different to the heteronormative spaces in which many LGBT people worked, went to school and at times lived (see Valentine 1996, Holt and Griffin 2003).

The majority of those who answered the CMIT questionnaire said that they enjoyed using or going to LGBT venues and events in Brighton and Hove (73 per cent, n. 591) and many spoke of having good nights out and enjoying the scene:

> Dave: I'd probably go to gay bars about 90 per cent [of the time], 10 per cent straight. On the gay scene, that's where I feel more comfortable. I'm not so comfortable when I'm in straight places, not 100 per cent, no. I think there are few places in the world that have so many choices for gay people as Brighton. I think for a good gay lifestyle, it's great. (Deaf focus group)

The scene can be a place where people found community and ease. Dave sees Brighton as offering choice available in 'few places in the world', and reiterates the difference of Brighton (see Chapter 3). Discourses about the possibility of 'being yourself' in gay spaces (Holt and Griffin 2003, Valentine and Skelton 2003) and expressions of sexualities persisted in the context of legislative equalities in the gay city, where 'we are everywhere' (see Chapter 3):

> Gay venues are much less judgemental and I can be my true self in them, as opposed to straight venues. (Questionnaire 68)

For others, being your 'true self' was possible in straight pubs and clubs, and some respondents described enjoying and feeling comfortable in non-scene clubs. The role of the scene in creating connections and community was not negated by seeing straight bars and mixed (gay/straight) social spaces in Brighton as 'safe' (Browne and Bakshi 2011). The gay scene created 'majorities' and made some LGBT people feel ordinary. This ordinariness was in part predicated on heteronormativities and homophobia that continued to be a part of LGBT lives in the gay capital (see Chapter 3):

> Sense of community. I'm comfortable going to straight venues, but enjoy feeling part of the majority. (Questionnaire 606)

> Natalie: I do still think homophobia's out there and it adds additional safety and security to be in an LGBT venue for a lot of people, compared to walking into a venue that's not LGBT specific, even if it's filled with lots of other LGBT people. There's still always that thought that something might happen or there might be people in there who aren't friendly.

> And I guess a way of connecting with people as well. If LGBT people are watered down around the city in various places, the chances of meeting up with each other is going to lessen. So having venues where you can go and meet other LGBT people is probably quite a good thing.

The scene in Brighton was characterised as different to straight spaces, and respondents asked for straight people to be kept out, resisting the heterosexualisation of gay spaces (see Binnie and Skeggs 2004, Casey 2004, Skeggs 1999). Thus, whilst the scene had commonalities with straight pubs and clubs, which were understood as welcoming, distinctions continued around the axis of straight/not straight. Moran et al. reported that Manchester's Gay Village was perceived by some as a place of safety 'against danger, which is always elsewhere' (2004: 161). In this research, safety, security and empowerment were experienced in gay scene spaces, in part by generating majorities and limiting straight invasions (see Skeggs 1999, Casey 2004). Straight 'invasions' could be read as making LGBT spaces normal, in heteronormative terms. Instead, reading the city as needing places where LGBT/gay and lesbian people are in a majority creates different form of ordinariness. An ordinariness in difference, offering 'friendliness', 'community' and 'feeling part of the majority' for LGBT people, who felt othered in straight spaces.

Some academics have included social groups and networks in their understandings of scene spaces (see Valentine and Skelton 2003). Alongside the proliferation of online friendship and dating networks (see Mowlabocus 2010) there were assertions that, in Brighton, community groups, including choirs and sports groups, accommodated those who did not fit or wish to socialise in commercial bars and clubs. These forms of social networking were understood as having different cultural expressions. Brighton as the 'gay capital of the world' offers diversity and 'something for everyone' and Harriet separates scene and community groups to this end:

> Harriet: Brighton is the gay capital of the world – one of. And it's got a diverse community. Whereas people in other areas of the country have just got a scene and that's about it, we've got quite a lot of work going on with different community groups. So you don't just come to Brighton and you have to go out on the scene and that's it. You can do LGBT sports groups, or youth groups, or this, that, and the other. You don't have to literally be involved in the scene to be a part of that community and feel like you belong. So I think it's got something for everyone, really. Rather than just those that want to go clubbing.

The public visibility and celebration of groups was related to broader societal and cultural contexts in Brighton, recreating the city as a 'safe' place. The urban idyll is lived and recreated through public and semi-private gatherings of LGBT people where they did not feel vulnerable to homophobic abuse. Indeed members of the Rainbow Chorus, an LGBT choir in Brighton and Hove, discussed the positive aspects of being involved both in singing and socialising with a group that respects and celebrates LGBT diversities (Watson and Goodhall 2010). Thus, the ordinariness of scene spaces extended to non-commercial LGBT social spaces and for some, these spaces addressed the inequitable hierarchies between LGBT people.

Even where respondents did not use the scene regularly or did not feel they 'needed' it for connections, safety or community, it held a symbolic importance. This was about knowing it was there rather than using the scene. This was part of a broader sense of place that saw Brighton as 'having it all'. Such place based imaginings also led to a belief that whatever LGBT people wanted, it could be easily created. This in turn led to the presumption that any absences in the scene could be filled 'without much effort':

> Brighton has a huge range of LGBT friendly or specific venues. I don't see that there's anything particularly missing, certainly nothing that couldn't be arranged by specific interest groups without much effort. (Questionnaire 264)

The imagining of Brighton as having a 'huge range of LGBT friendly venues', as well as the belief that everyone can 'fit in' and enjoy the scene and that 'something' can be arranged by 'specific interest groups', points to particular forms of privilege (see Noble 2012). As we will discuss, this was contested by other respondents. It also indicates the normalisation of some LGBT people in the city, such that they are part of the possibilities that gay Brighton affords through adhering to particular conditionalities of inclusion (Richardson and Monro 2012).

As we move to explore problematic elements of the gay scene in Brighton, we hold in tension the otherings of scene spaces with ways in which these spaces did more than exclude. We follow Bassi (2006: 226) in seeing the commercial gay scene as more than 'simply a spatially contained commodification of difference incorporated within capitalist hegemony'; Brighton's gay scene does offer the possibilities for resistance, contestation and inclusions, fracturing and fissuring the hegemonic classed and gendered scene. More work is needed to explore these spaces of possibilities alongside critiques of normalisations. Indeed, we would argue that the desire to be included and to be 'part of it' was a testament to the importance of the gay scene in being part of the gay city.

'For Me There is No Scene': Marginalisations in and from Gay Scenes

Belonging to the scene was a key aspect of being part of the gay community in Brighton. Not belonging and being rejected by the scene was particularly painful not only because it implicated the group that should accept you, but also because it disenfranchised some LGBT people from the promise of this city. It has been argued that scene spaces are experienced unequally, affected by intersectionalities including race, age, class, bodily ideals, gender identities and sexual identities (Bassi 2006, Binnie and Skeggs 2004, Casey 2007, Doan 2009, Faderman 1992, Hines 2007, Holt and Griffin 2003, Nash 2010, Taylor 2007, 2007b, Taylor et al., 2010). In this section we delve into who, ideologically, made up the scene, before focusing on those who did not fit imaginings and manifestations of scene spaces.

Scene spaces can be associated with a particular form of 'acceptable gay': those who fit particular norms and work within structures of appropriateness and decency (see for example, Bell and Binnie 2000; Richardson 2005). In Brighton, pub and club environments accommodated certain (gendered, raced, aged and classed, as well as music) cultures, which were then attached to the label 'gay', recreating what could be gay. The expectations of what Brighton *should* offer to all LGBT people were in part built on scene spaces, and were unfulfilled for some:

> Jo: What do we think of when we think of Brighton and Hove? People come for the scene, don't they? St James's Street is known nationally – worldwide isn't it? – as being part of the gay community. Brighton's seen as the gay capital of Europe. So it is very clearly that's what it's about and everything domino effects from that really.
>
> Because people come here for the scene, they have a huge influence. Then the impact of those people coming down here for the relative safety of the scene,

of the community ideal of what the scene brings to people, then they have a huge impact on the rest of Brighton and Hove around health services, housing services and the rest of it. Cos people get here, they engage with the scene and they want to stay.

I think the scene is seen as a safe space in regards to community and identity, but I think it's incredibly abusive. We know from studies that have been done. People use the scene to get a bed for the night. There are issues around visibility because it is about this area of the city and that's where you can find them. I know that people come up on the train looking to gay-bash somebody. So I think there are real issues.

For some, feelings of culturally belonging in the city were created through engagements with the scene and this had impacts for services, such as housing. Jo reiterates the vulnerabilities around migratory trajectories informed by Brighton as it is constructed through experiences of the scene (see Chapter 3, Myslik 1996). It is clear then that despite the idealisation of the scene and Brighton's LGBT social spaces, and contrary to the assertion in questionnaire 264 above, not all LGBT people had the resources (money, knowledge, connections, expertise, confidence) to go somewhere or 'put on' something they would like to see (and even this presumes a possibility of somewhere you might 'fit').

Feeling excluded from the scene was central to the marginalisation and, at times, isolation that characterised some LGBT people's experience of Brighton. Not fitting in, in either gay or straight spaces, resulted from, and in, being marked as 'other' and 'out of place', even where spaces were purported to be accepting and inclusive. The label 'gay' scene and the places and practices that informed it were not inclusive of the range of identities and practices encompassed in LGBT collectives. In the questionnaire data, less than half (42 per cent, n. 25) of trans people agreed that they enjoyed using or going to LGBT venues and events, compared to 74 per cent (n. 562) of non-trans participants. Lesbians/gay women (68 per cent, n. 188), bisexuals (64 per cent, n. 17) and those who defined with other sexual identities[2] (56 per cent, n. 27) agreed with the statement less frequently than gay men (79 per cent, n. 339). Although this shows that the majority of bisexuals, lesbians/gay women and those who defined with other sexual identities do enjoy the scene, there were also significant levels of disenfranchisement within these groupings. Indeed, where 'gay' was associated with male, body beautiful culture, dance music, hedonism and youthfulness, then the creation of the gay scene through particular commercial, embodied, class, gender, racial, ethnic and sexual norms was apparent:

---

2   This is a collated category for the purposes of valid statistical analysis. This group includes anyone who defined as queer, questioning, unsure, heterosexual/straight and other.

Lee: Some of the more overt power is concentrated in the hands of men, and generally on the LGBT scene, so all the venue owners, pretty much, are guys. In terms of their attitude towards women and bi and trans people and lots of other things, that's quite telling. I think white, middle class, gay guys generally have power on the scene. I'm talking about people around the scene and just invisible things about the way that things are set up for them, in terms of things that are available – the entertainment, dodgy drag acts, the music they play, the kind of environment it is, the door staff, everything about it, the advertising images they use, the kind of cultural norms within the LGBT scene. In terms of male gay bar identity, I think that's quite an odd kind of space, purely because if you look at all the issues that get addressed in terms of LGBT, activists and workers agendas, a lot of them are about men. So you've got public sex environments, you've got people being attacked in the streets, obviously [both are] important issues. Both lean towards being male issues.

The 'homonormative gay male spaces' and privileged gay men were seen as enforcing class, gender and other norms in this research (see also Doan 2009, Nast 2002, Taylor 2007). Gay male norms were simply 'there' and this was coupled with gay men being both catered for by and representative of 'the LGBT scene'.[3] Yet, Lee also notes the importance of the scene beyond hedonism, in the mutual formation of 'LGBT' issues and activisms, with prominent agendas and work undertaken under the guise of LGBT, but with the majority of it focusing on men. In this way, city processes in gay Brighton that made some LGBT people ordinary were in part created through scene spaces. Lee went on to point to his 'place' in the scene and the ordinariness this affords him, identifying 'others' who he felt were excluded:

Lee: I actually quite like the term gay. It feels a bit like a dirty word though. It always feels quite naughty because I know that it's extremely un-inclusive in lots of ways. Yet my identity kind of jumps around between being gay if I'm on a night out and being quite naughty, or LGBT if I'm being 'at work' or queer if I'm being politically active or different other stuff; or, just not straight. I think all of them have got uses.

Generally if I hear the word gay it kind of puts my hackles up because I use it [in] a semi-ironic way, a playful way, but generally when it's used it's in quite an unthinking way. With that said, generally if some organisation is using the term LGBT, it can effectively mean gay, it's kind of shorthand for it. Similarly, they can use the term gay and they hope it means LGBT, whereas it doesn't at all.

---

3   As we have argued elsewhere (Browne and Bakshi 2011), socialising in Brighton was not restricted to gay spaces even for those who 'fit' social norms, particularly in terms of class and ethnicity.

Lee's use of the term LGBT and gay was spatially specific, disruptive and 'naughty'. He deployed the term 'gay' ironically in scene spaces, indicating his acceptance in these spaces. Recognition of the spatial differentiations of LGBT work and 'naughty' gay playfulness illustrates the separation of LGBT services or communities and scene spaces (a point we explore further at the end of this chapter). Lee's narrative embodies the tension of gay Brighton, in being both 'naughty', and catering for vulnerable LGBT people who were not included in the category 'gay' and were not understood as part of the gay scene. In this research differential symbolic and material access to scene spaces was apparent in relation to sexual identities, ethnicity, D/deafness, age, mental health issues, class and practices such as drinking and taking drugs. We will now examine three of these that were discussed in nuanced ways by participants: gender, class and drinking. Through these discussions we offer insights into the reconstitution of scene spaces across differences that include but are not limited to identities. This has implications not only for elucidating the diversity between LGBT people in relation to scene spaces, but also who is part of the ordinariness the gay capital affords through scene spaces.

'Gay Bar'

Gender has been a significant factor in the appropriation, use and visibilities of lesbians and gay men in urban spaces and this is often related to the visibility and control of scenes. Whilst gay men are seen to appropriate territories creating highly visible 'ghettos' through shops, bars and clubs, it is argued that women use space in different ways, gaining 'territories' through where they live, social networks and known signs that make lesbian bodies (in)visible (see for example Adler and Brennan 1992, Podmore 2001, 2013, Pritchard et al. 2002, Rothenburg 1995, Skeggs 1999, Taylor 2007b, Valentine 1993). Pritchard et al.'s (2002) study of the 'Vanilla' bar in Manchester points to the difficulties experienced by lesbians in claiming space in a mainly gay male scene that does not 'particularly welcome women' (105). Yet discussions of homonormativities and the normalisations of both 'lesbians and gay men' can overlook the centrality of gender, both in terms of diverse masculinities (see Elder 2004) and the differences between women. In this research gender was a significant axis of difference, which meant that some lesbians[4] felt excluded from both heterosexual cultures and gay scenes – what we have termed multiple marginalisations. The multiple axes of gendered and sexual differences also included the provision, availability and accessibility of spaces for trans people and bi people in contrast to gay men and non-trans people. In

---

4   We use the term lesbian here to recognise the longstanding differences between women who use the scene, including 'straight' women who use the scene with gay male friends, sometimes called 'faghags' (see Casey 2004, Skeggs 1999, Pritchard et al. 2002). Arguably certain forms of femininity (among other things) can operate to exclude some women, lesbian or not (see Taylor 2007b).

this chapter, we focus on women's experiences, focusing mainly on lesbians, yet recognising that both trans women and bi women can experience similar issues pertaining to the non-trans/cisgendered[5] and male gendering of scene spaces (see Persia's, a trans woman, response at the start of this chapter).

The gay scene in Brighton was seen as constructed and catering for specific forms of (youth based) masculinities.[6] Although bars, clubs and pubs were supposedly 'mixed' in terms of gender, some women understood the scene as 'phallocentric' and generating 'male' vibes:

> Natalie: My experience is there's not much around for women. [It's] difficult going into gay venues that don't necessarily market themselves as gay men only but there's that vibe. That's kind of an uncomfortable feeling, to have the sense that men don't want women there.

> It amazes me at Pride, the number of lesbians that you see, and I just wonder where those lesbians socialise the rest of the year. My feeling is they tend to keep themselves to themselves in their own small social groups and perhaps mix and mingle in non-LGBT venues, or in each other's house[s]. Which is great, but I think the downside is it becomes a group that you can't permeate because where are these people? How do you find them? How do you make connections? Maybe for people that are a bit more socially adept, they can do that, but I wonder about people that need a bit of help. How do they break through into the lesbian world in Brighton when there isn't anything set up to help them do that? Which makes me think there must be an awful lot of marginalised lesbian women.

We have argued elsewhere (Browne and Bakshi 2011) that LGBT socialising and leisure was not confined to scene spaces. Here Natalie reads this as recreating the masculinities of scene spaces, created in part through 'vibes', entertainment, environments and cultural norms (see also Lee above). She and other participants noted the lack of publicly visible and accessible lesbian orientated social spaces that enabled particular LGBT people to be seen, known and connected. More than this, as we have seen, such spaces recreate the 'shop window' of gay Brighton,

---

5   Cisgender is used here to refer to non-trans people. It names the privilege of the alignment of primary and secondary sexual characteristics and gender identities. This is usually taken for granted where for example the alignment of sex organs, male identities and masculine practices is seen as natural or normal. Cisgender challenges the othering of trans people, refusing to see men as 'non-trans men' and women as 'non-trans women'. Instead it argues that, where this is meant, cisgender man or cisgender woman is used to qualify the term. Thus, where man or woman is used this automatically includes trans people who identify with these genders/sexes.

6   See also Valentine and Skelton 2003 for a discussion of the problems and tensions experienced by young people on gay scenes in the North of England.

illustrating the symbolic as well as material exclusions of the gay city. The desire to create lesbian/women's spaces however may not be commercially driven and some did not see the scene as responsible for, or able to, create inclusive women's space:

> Lesbians, especially older ones, have a long tradition of networking and socialising in small groups – we maybe don't need a club or a bar – but lesbian cultural pursuits would be most welcome. (Questionnaire 830)

Some lesbians may not want to be a part of gay scenes and spaces, contesting understandings of these spaces as exclusionary and marginalising. Indeed a focus on the gay scene as a locus of community and belonging afforded it power that some women questioned. This did not negate male power and control of scene spaces (see also Pritchard et al. 2002), but does refuse to relegate all lesbians/LGBT women who did not use the scene to the position of victim. Other forms of 'lesbian cultural pursuits' suggest different 'scenes', spaces and possibilities that may not be territorialised through the visible ownership and existence of bars, pubs, clubs and shops (Podmore 2001, Rothenburg 1995, Valentine 1993).

Yet, we are wary of seeing community groups as the saviours of the scene. These social networks and groups were also experienced as marginalising by some. For Kali, her ethnicity meant that she felt her experiences were different from the white dominated lesbian groups in the city:

> Kali: I wouldn't go along to the Women's Walking Group, although I'm a woman and I like walking. What tempers my enthusiasm is experiences all through my life of going to something like that and not fitting [in]. It's not that I look different and people are horrible to me, it's the sort of things that people talk about and their attitudes to life are very different to me. Life experience has told me that probably I won't have anything in common with them and it won't lead to friendships, and really it is quite a negative experience for me to feel like they're all having a good time and chatting making friends and [that] I can't be part of that, something that fits with one layer [of my identity] isn't going to fit with the other layers. (Pilot focus group)

Acknowledging these multiple and intersectional forms of othering can mean that marginalisations from the gay capital through exclusions from LGBT spaces cannot be solely attributed to scene spaces. Experiencing being part of Brighton was predicated on 'layers' of identities that don't 'fit' in intersectional ways that have long been discussed by Black feminists (see Appendix 2).

Brighton's perceived lack of lesbian- and women-focused venues and events was often compared with ideas about what was provided for men. The illusion of gendered capital for all men who supposedly 'fit' the category of male in 'male gay spaces' was contested by those who did not conform to the figure of the 'acceptable gay' (Bell and Binnie 2000). The symbolic creation of gay male identities through scene spaces and icons can mean that for some, the conflation of gay scene with LGBT identity and community was alienating. Although at times related to intersectional difference pertaining to race as Kali discussed and class to which we now turn, this was not always the case. Again, it is important to note how LGBT people were not only disenfranchised and excluded but they also created other spaces and modes of doing and being 'who I am':

> Simon: When I came out, the presence of the gay scene as the only visible image of what it was to be gay was quite harmful to me and delayed my coming out. [An] important breakthrough for me was to say that's not my scene, that's not for me. I don't have to subscribe to everything that happens there. My journey to being who I am was through other political movements: the women's movement and peace movement. I'm quite lucky to have had a different model than a kind of male commercial scene, which has inspired me. I run around being inspired and hopefully inspiring other people, there's lots of issues that are not heard in our community and I'm very passionate about them. (General focus group one)

Economies of Exclusion?: More than Money

> Brighton's working class [are] forever treated not as an essential part of Brighton and Hove's much trumpeted 'diversity' but rather as a dirty great Burberry[7] clad burden on the sleek, mobile, modern clean machine that is 'Brighton and Hove'. (Burchill and Raven 2007: 1)

Scholars have shown how issues of class intersect with sexual and 'gendered resources and entitlements' (Taylor 2007: 173, 2007b, 2009, McDermott 2011). Class is manifest in geographically specific ways such that class distinctions are made through spaces, as well as remaking them (see Binnie 2011, Browne 2011).[8] Burchill's quote points to the classed ways in which Brighton's 'diversity' operates reflecting the 'trendiness' of Brighton's symbolic and material, classed location in the 'affluent' south-east of England (see Chapter 1). Social and cultural economies of class on commercial gay scenes, although related to money, encompass more than material resources. Scene spaces are important to the reconstitution not only of city imaginaries of Brighton as the gay capital, but also to LGBT lives in the

---

7   Burberry is a reference to a particular clothing label that is seen to link with the often maligned 'chav' culture (see Jones 2011). The use of this term implies an association with a hostile and undesirable working class identity and culture, see also Taylor 2007b.

8   Fuller and Geddes (2008: 262) argue that under New Labour in the UK the emphasis on 'community' moved class away from broader economic and political processes to instead 'reduce social inequalities to the deprivation of socially constructed poor neighbourhoods'. Our focus on neighbourhoods reflects this agenda, as well as being grounded in the cultural understandings of particular areas and the material – particularly housing provision – geographies of class.

city. Thus if, as Taylor (2007b) shows, scene spaces can enforce particular norms of femininity or class manifestations through appearance and 'proper gay' signifiers, class plays a crucial role in how Brighton constructs LGBT lives. However, this is complex. As with most gay villages, Kemptown has been, and remained at the time of this research, predominantly resided by working class people, with rising house ownership and gentrification coexisting with social housing (Laurie and Knopp 1985, Knopp 1990, Bell and Jayne 2004, Hubbard 2004). It also bordered the working class area of Whitehawk. The tensions between middle class gay gentrifiers and traditional working class communities set the context of this research, and the 'cleaning up' of Kemptown was a key driver for the move to label this area the gay village.

In this research, class was at times associated with appearance, as others have noted. For example the wearing of 'Burberry' was mentioned as a marker of difference and a term of abuse illustrating the cultural as well as material aspects of class (see Taylor 2007b, Binnie 2011) and also residential areas, which we now focus on. Areas of social deprivation that were seen as 'dangerous' were often portrayed as inherently 'anti-gay' (see also Taylor 2009). These areas were feared and denounced by some participants as places where hate crime and a lack of acceptance were located there. This was seen as an exception to the overall acceptance of gay Brighton, pointing to the intersections of class, place and sexual identities (see Taylor 2009). These intersections gave rise to multiple marginalisations, where some felt othered both by LGBT people and those in residential communities. However, identities and communities related to class and those related to gender and sexual identity were not mutually exclusive, but they did at times exist in tension. Memberships of neighbourhood communities were negotiated with gay lives and practices:

> Michael: I don't know if I would feel safe being in St James's Street when I know that someone from round the corner could come down and recognise me going into a pub. Everything just seems so close together, whereas you can just pop into London and then go back out to Wembley or somewhere and no one will ever know.

> Ed: I was on a seat in [name of public sex site] and a few boys came up to me and I thought 'these have got to be Whitehawk boys'. So I pretended to be asleep and they started shouting at me and I said 'Will you just fuck off, I'm trying to have a bit of sleep here, I've got pissed and I'm trying to sleep it off'. [He said] 'Oh you're one of them fucking bum bandits!' and I said 'What are you talking about? Can you see anybody?'. He says, 'No, but they're coming' [and] I said 'Oh go away will ya?'. They were just about to do it and this boy came from nowhere and said 'Go on, leave him, he's all right', and I thought 'Oh, where did you come from?'. I know who the boy is.

> Maybe it's because my face [is] known and I'm known in Whitehawk, do you
> know what I mean? Not, thank God, not as [name of drag persona] but I am
> known as me. (Pride in Whitehawk focus group)

The use of gay spaces not only enables an enactment of certain sexual and gender
identities, it also marks people who use such spaces as 'gay' or LGBT (see for
example Bell et al. 1994, Browne 2006a, Levine 1979, Mowlabocus 2010). As a
small city, Brighton did not offer the anonymity which was perceived to protect
those in London from being known as LGBT to their neighbours and potentially
suffering abuse in their neighbourhoods. The city formed LGBT lives, but not in
uniform ways, and there were complex interactions with specific city classed or gay
spaces. For Ed, 'being known' in Whitehawk had protective possibilities, as well
as dangerous moments. 'Being known' protected him in a public sex environment
where he could deny interaction with the space but entering or leaving bars in St.
James's Street leaves little room for denial. Ed speaks of averting the immediate
threat because of where he lives and where he was sitting, whereas Michael
notes the lasting implications and vulnerability he would feel if marked as 'gay'
through the attachment of scene spaces to identities. Thus, not only was there a
differentiation between LGBT, there was a spatial negotiation between fitting into
Whitehawk and accessing the gay scene.

Class was also a basis for exclusion from the gay scene. As a manifestation
of multiple marginalisation LGBT people who lived in working class estates felt
marginalised not only in the neighbourhoods in which they lived (see Browne
and Davis 2008), but also by the negative reactions from other LGBT people *in
response to where they lived*. In focus groups, participants spoke of the way they
dealt with this form of prejudice:

> John: I have people [on the scene] say 'I'm not coming back with you because
> you live at Whitehawk'. So I say I live at East Brighton. (Pride in Whitehawk
> focus group)

This tactic was also reflected in the quantitative data which found that 21 per cent
(n. 72) of those who lived in working class areas did not tell people on the scene
where they lived.

Negotiating belonging and exclusion occurred in the symbolic as well as the
material realm. Even where respondents did not tell people on the scene where
they lived and said that they could not afford to use scene spaces, the importance
of belonging on the scene, and thus being a part of gay Brighton, was apparent:

> Ed: I just go out when I feel like it or when I've got the money for it, which
> isn't much. Just go out to various clubs and pubs and just meet people and have
> a good time.

John: Mine's pretty much the same. I only go out when I've got money or if someone asks me, even then when I go out I barely drink. That's about it.

Researcher: So is it a problem if you don't have money?

John: Well nine times out of ten I haven't got any money.

Researcher: Are there enough events and activities that are relevant to you, or that you can afford?

Ed: Well, there's something going on all the time, somebody performing somewhere.

John: Every night of the week.

Ed: You can go and do something. [name of club outside of Kemptown in a working class area of the city centre] always has something on. You can go to the piano bars and there's performances of people at various theatres, [names], to see people perform. Yeah, I'm sure there's loads, something on every night for people if you want to go. It's people that make the effort to go, they don't usually bother or they've not got the money, which is an obvious one. (Pride in Whitehawk focus group)

Ed and John clearly wanted to convey to the researchers their knowledge of and inclusion in the scene in Brighton, aligning themselves with the gay capital and perhaps demonstrating their gay cultural capital (see Bourdieu 1986, Taylor 2007b). This not only located them as part of gay Brighton, the place of Brighton recreated their (sexual) identities. This was the case even where there were dissonances between these alignments and their enactment of LGBT lives. In this focus group discussion moved between the 'I' who has no money and the imagining of a scene that always 'has something going on' that 'you' can go to 'if you want to go'. Taylor (2007b: 161) notes the 'fraught and fragmentary' ways that working class lesbians make entitlement claims on the scene. Here gay men from working class areas discuss their sense that they were 'part of' the scene, illustrating the symbolic and significant role of the scene beyond those who used it. Such narratives also pointed to the potential inappropriateness of naming some LGBT people as 'marginalised', when they did not (wish to) see or name such exclusions – or at least not to researchers looking at LGBT lives in Brighton. More than this, aligning with gay Brighton allowed the creation of identities that were 'sorted', 'fine' and 'accepted', indeed ordinary, despite the classism that they encountered on the scene and the fear that some expressed of being associated with scene spaces where they lived.

Drinking and Belonging

Focusing on identities and intersectionalities, such as gender and class, can lead to the overlooking of important practices that create scene spaces. Brown (2009) notes the importance of knowing sexual signs and cues in using and fitting into cruising spaces, similarly practices such as drinking and drug taking can be key to who was 'in place' in scene spaces. In Brighton, bars and nightclubs dominated gay commercial scene spaces. Drug and alcohol use amongst LGBT people has been the subject of much medical and increasingly community research, predominantly working from an understanding of drinking and drug taking as 'self-destructive' (Valentine and Skelton 2003: 582, Adams et al. 2008). Conversely, there has been a dearth of investigations that focus on those who *do not* drink or take drugs.

Research has identified that drinking spaces are more than simply sites of disruption and potential/actual violence, and are instead assemblages of social relations that reconstitute everyday lives, as well as having positive feelings and emotions attached to them (Jayne et al. 2010). As Jayne et al. (2010, 2011) argue, alcohol and drunkenness offer positive and enjoyable experiences that are often not captured in the policy debates or moral panics that surround the regulation and surveillance of urban centres. Thus, not drinking may have more significance for social inclusion, belonging and wellbeing for LGBT people than has been addressed to date. Whilst drinking spaces within scenes did not constitute the only (or even the predominant way) that LGBT people socialised (see Browne and Bakshi 2011), this chapter has shown the ideological, as well as the material, importance of scene spaces to being a part of gay Brighton.

In this research 85 per cent (n. 669) of respondents said that they drank alcohol and 50 per cent (n. 406) said that they had taken illegal drugs or used legal drugs without a prescription in the past five years[9] (see Browne, McGlynn and Lim 2009). 77 per cent (n. 510) of respondents who drank alcohol agreed with the statement 'I enjoy using/going to the LGBT commercial venues and events in Brighton and Hove' compared with 50 per cent of those who did not drink alcohol (n. 59, p<.05, see Table 4.1). 82 per cent (n. 330) of those who had used drugs agreed with the statement 'I enjoy using/going to the LGBT commercial venues and events in Brighton and Hove' (see also Buckland 2002, Halkitis et al. 2007, Kipke et al. 2007, Valentine and Skelton 2003). They were more likely to do so than those who had not taken drugs (63 per cent, n. 253, p<.05). In other words, those who drank and took drugs were more likely to enjoy using scene spaces.

Linking LGBT communities to gay scenes in commercial venues that are focused on the sale of alcohol perpetuates a very particular form of gay/LGBT culture that values drinking[10] (and, perhaps to a lesser extent, drug taking).

---

9   We used 'taken drugs' as short hand for this category.

10   This could be seen as a general 'Brighton'/post-industrial city trait, as a space and culture that values drinking and gains significant revenue from night time economies (see Chatterton and Hollands 2003).

Table 4.1 Enjoyment of using/going to LGBT commercial venues/events in Brighton & Hove, by alcohol use

		I enjoy using/going to LGBT events and venues in Brighton and Hove				
		Agree	Disagree	I don't use	Unsure	Total
Drink alcohol	No.	510	38	68	50	666
	%	76.6	5.7	10.2	7.5	100
Don't drink alcohol	No.	59	15	31	13	118
	%	50	12.7	26.3	11	100
Total	No.	569	53	99	63	784
	%	72.6	6.8	12.6	8	100

Alongside the documented ways in which drugs and alcohol can be used to 'cope' with discrimination and abuse (see for example Weber 2008, McCabe et al. 2010), participants pointed to a lack of alternatives to the scene for gay/LGBT socialising and meeting new people and potential sexual/relationship partners:

> Shane: That's a lot of pills to be popping that weekend and that's a lot of alcohol to be drinking and actually in part I think it's because there's nothing else to do. I mean really, how else are you gonna meet people, where else are you gonna hang out with people? (First generation immigrant focus group)

> Lee: I think if you want some kind of LGBT belonging and you don't have a massively well-developed LGBT friendship network – that can be the only place where you can be LGBT, at the venues. Of course what goes along with that is drinking.

Perhaps related to the place of the scene in developing LGBT 'communities', social networks and support systems, those who did not drink were more likely to have experienced domestic violence, to have had mental health difficulties and have had serious thoughts of suicide, and were likely to feel less safe, according to the CMIT data. Although alcohol use was high across all categories, some groups were statistically more likely to drink alcohol, and these were groupings which could be read as more likely to 'fit' in the scene, including gay male respondents, men, white respondents, young respondents, non-trans respondents and those who earn more (for full details see Browne, McGlynn and Lim 2009). Thus, in a place where the gay scene holds significant social and cultural power, not drinking alcohol was associated with isolation and marginalisation:

> Gabriele: I don't drink, although I do go to pubs. I don't go on the scene clubbing or anything like that, and I'm trying to sort of increase my circle of friends. It's sort of, how do you meet people? (General focus group 2)

> Maria: I used to go out a lot on the scene but not so much over the last couple of years. I became sort of quite isolated at home. (Pride in Whitehawk)

This was supported by the quantitative data in the CMIT research. Those who said that they feel isolated in Brighton and Hove were less likely to drink than those who did not feel isolated. From the statistical data it is not possible to know whether or not drinking led to not using the scene, or whether marginalisation from the scene resulted from not drinking. Nevertheless, avoiding drinking might mean avoiding spaces which revolve around this activity. Respondents noted how not drinking was not respected on the scene:

> Jude: I don't drink. Sometimes when you go out and about and say 'Can I have an orange and soda or even anything else that is non-alcoholic' they kind of look at you as if to say 'What?' [laughter]. You're out on a Saturday night and they're thinking 'What, not drinking?' A lot of places seem to think that if you go out you must drink and I think a lot of people don't understand that some people don't drink.

> Phil: Yeah, it's an unusual stigma with not drinking and it's really bizarre to me because I'd be praised for not doing something which is potentially harmful. (Young people's focus group)

Jude and Phil were seen as out of place, not because of a recognisable social difference such as age, but because of their failure to 'fit in' in terms of drinking practices. Such findings suggest that exclusion did not follow easy lines of identifications and practicing cities in particular ways also led to isolation and othering, even amongst those who supposedly occupy homonormative positions (in this case young, white men). Alcohol was often understood in terms of health needs and particularly the embodied 'negative effects' of drinking. Phil contended that such health considerations were elided by the social pressures and 'stigma' of not drinking.

What Role Should the Gay Scene Play?: Social Service or Commercial Entity?

It is clear from the previous section that fitting into LGBT spaces was differentially experienced by LGBT people all of whom in theory *should* have a claim on the gay capital through these spaces.[11] As we have seen, some women spoke of spaces where they felt they fit, and there was an active contestation of the exclusions that were read as inherent to the gay scene in Brighton. The gay businesses and those who frequented them, as apparent in the Wild Fruit quote above, were seen as both separate from 'the community' and at times deployed as 'the gay community'.[12] Those who 'join us for a party and support your community' can 'simply lounge', without addressing the exclusions and otherings within and from these spaces explored in this section. Thus, the perception of many of our participants was that gay businesses in Brighton had symbolic and material forms of power that drew on particular logics that separated commercialism from community, such that this was not the 'whole city' for them. Having explored experiences on the scene, we now turn to examine the tensions of commercialisation and community, where the consumption of commercial scenes can be seen as constraining (as well as enabling) (Bassi 2006, Holt and Griffin 2003).

Knopp (1992) argues that from early on, gay personal lives and sex were commoditised (see also Bassi 2006), and D'Emilio (1983) contends that capitalism created gay and lesbian lives and communities. However, during the early twenty-first century the community and voluntary sector and the scene were more explicitly and overtly de-coupled in gay Brighton.[13] Community and support functions previously attributed to the scene, such as providing safe space for LGBT communities to come together, were increasingly located in the voluntary and statutory sectors and away from market forces. The scene was then read by some as solely market driven yet, as we have seen, the gay scene held a significant role in creating gay Brighton and thus reformed LGBT lives. We explore the diverse and tense relationships between community and scene and see place as active in the reconstitution of the materialities and symbolic power of both communities and gay scenes.

Speaking *as* the scene or *for* the scene was construed and portrayed as speaking *as* or *for* gay/LGBT people and communities in Brighton. Yet, such

---

11   Further work (again both in terms of social support and academic endeavours) is also needed to explore these and other multiple axes of social difference in gay cities, for example, age. Our data suggests that older people were more likely not to use scene spaces and ask for something other than what was currently provided (see Browne, McGlynn and Lim 2010, Browne, Bakshi and Lim 2011). However, some younger people also felt that the scene did not cater for them (see Browne 2007, Valentine and Skelton 2003).

12   In the Wild Fruit publicity, 'the community' is portrayed as both part of the party, and simultaneously 'community issues' are distanced from 'the party'. The use of the term 'community' by a commercial event is not unusual, given the importance of these spaces in creating visibilities and (fleeting?) collectives. Bassi (2006) notes the ways in which the 'social wealth of Birmingham's gay scene' was reiterated in terms of the 'community' ownership of the Nightingale night club, following the failed attempt to take over this venue by a gay entrepreneur.

13   In Brighton the majority of LGBT community and voluntary groups were established after 2000.

representations were dubious. There was little commonality, agreement or joint working amongst the gay business that supposedly made up 'the' scene. Indeed, gay businesses in Brighton rarely operated in collective terms and worked together only sporadically. More often gay businesses competed with each other and this 'community' was seen as doing damage: 'backstabbing each other and slagging each other off' (Robert). In-fighting amongst these businesses did not negate their assumed influence on the socialising, drinking (and drug taking) practices of LGBT people, or their power in reaching a large number of people and in speaking for the 'LGBT community'. Neither did it affect their symbolic role in creating gay Brighton. Understanding the scene as united, 'the gay community' created dubious coherence as well as doubtful association with gay/LGBT collectives city-wide. Some noted that the ownership of most gay businesses did not lie with LGBT people. Rather, the scene was owned by 'straight' breweries. Yet, the presumption of commonality and links between gay businesses (mainly bars and clubs, but often including the local gay press, which presents a unified scene in part through its mapping and advertising, see figure one) necessarily drew on *one* shared sexual identity and some element of shared histories/responsibilities for 'gay money', as well as competing with each other for it:

> Gordon: I see a big failure of so much of what we do is about not being able to explain to the business sector in Brighton what their responsibilities are. So I would say that anybody who's looking to move the work forward, they would have to take on the business sector and say to them 'You're in this town because it's a gay town, to make gay money and there's a responsibility that comes with that'. (Consultation event, April 2009)

Gordon's remarks point to the importance of business for 'anyone' (read as the community and voluntary sector) looking to 'move forward'. A trajectory of progress is invoked and the assertion that Brighton can do better is seen as an impetus for moving 'the work forward'. More was expected of the gay scene in Brighton beyond taking 'gay money', and place was active in placing 'responsibilities' with business owners. Many of our participants subscribed to the ideal that 'they' (gay business) should cater for 'us'. Beyond fundraising then, businesses were deployed as 'the community' who educate, and were educated, to provide for the voluntary sector:

> Joseph: I think some of the businesses wouldn't know if they were marginalising people. I really do. My experience of the business sector is that they work in this bubble. Unless the voluntary sector goes to them and educates them, they don't understand. The greatest supporters and helpers to me are people that I've educated and taken and explained how the voluntary sector works.

The responsibilities associated with new (neo-liberal) sexual citizenships (see Richardson and Monro 2012, Stychin 2003, 2006) are apparent here. It is the job of those who feel excluded or who support marginalised LGBT people to 'educate' and in this way obtain support, both financial and resources. Conversely gay businesses were expected to 'give back'. Some scene venues and event organisers were, to an extent, aware (or taught) about their responsibilities associated with receiving and seeking 'gay money', and these invested in being part of the LGBT/ gay collective. This was often through giving money and space to community groups (see Aeon events, 2010 that started this chapter). Creating a 'we' comprising all gay/LGBT people relies on a division between straight/non-straight, where straight spaces pathologise and deprecate non-straight identity. This creates a specific form of ordinariness, where 'gay community' is itself ordinary without necessarily referring to heteronormative orders or straight inclusions. However, the ways in which this ordinariness was created normalised specific gay men and overlooked the multiple ways in which scene spaces (as well as social networks) can create LGBT lives through exclusions as well as inclusions.

It is clear from these narratives that bars and commercial interests did not have to be invested in community work, and Brighton was unlike 'other places' where LGBT support and services were perhaps limited to and located in the gay bar. Thus, the scene and the community and voluntary sector were seen as interacting as well as at times existing in tension with each other. This division contests that 'a' gay community is created through scene spaces. The professionalisation of community support away from gay bars to LGBT specific groups,[14] as well as a public sector remit to provide for LGBT communities in Brighton (see Chapters 6 and 7), offers a different understanding to the ways in which LGBT non-profits operate elsewhere (such as in the USA). Their positioning as part of gay Brighton, yet separate from the scene, meant that some community and voluntary sector groups enacted contradictory roles:

> Natalie: Professionally I don't really feel that there's a great connection between those of us who are in the business of providing services like help, support, counselling and so on, there's no bridge between that and the scene. And maybe there can't be, because the two remits are so entirely at odds. They're about making money and finding a good time for people and we're about picking up the pieces and supporting people. But I do think we could work better together. I guess our only connection with them really is in pleading with them to do fundraising for us. Some of them are really supportive with fundraising. It must be difficult for them if one of our agendas is 'let's cut down the drug and alcohol use in the community', the last thing the pub owners want is notices telling people to stop drinking. So it is a challenge.

---

14   This is arguably reversed for sexual health organisations that understand the scene as a key space in which to offer support and interventions. The tensions between these organisations and the commercial imperatives of the scene are beyond the scope of this book, suffice to note that in Brighton and elsewhere 'strong' sexual health messages have been mediated through concerns of bar owners and others about 'scaring away clients'.

Natalie questions interactions between community and commercial sectors in the vein of the Wild Fruit quote that began this chapter ('you can show your support … just by coming along'). There were clear tensions arising from the different values held by scene managers and promoters and the LGBT community and voluntary sector. These were the constituencies that supposedly composed 'the gay community', which was polarised from straight identity. Yet here community groups held the role, often paid for by public sector funding, of 'picking up the pieces and supporting people' after the party (run by gay businesses that supposedly supported them in other ways). The relationship between commercial and community was thus far more complex than clubs and bars fundraising for 'good community causes'. At times the tense relations were based in not just the funding of services that 'pick up the pieces' for business, but also included an LGBT voluntary sector that was supported by public funding. The latter sometimes necessitated critiquing and working against the commercial operation of the scene in order to cater for LGBT people's health and wellbeing.

Divisions between commercial and community enabled some gay men to facilitate interactions and act in gatekeeping roles. They deployed the trope of 'gay community' to influence and control decision-making in the city. Such gay men influenced which groups were heard and supported both by scene businesses and in the media. Falling out with these people meant that community groups were subject to negative press and individuals were targeted and 'blamed'. The control exerted by gay men over charities such as Pride, which we discuss in Chapter 8, was closely tied to scene spaces. Consequently, in considering the creation of gay community through the scene, the hierarchies in place not only serve broad commercial interests, but they can also maintain the dominance of certain gay men.

In questioning these power relations some saw businesses as only commercial and profit driven. This meant that only certain aspects of LGBT lives were profitable and marketable (Binnie and Skeggs 2004) and removed the responsibility of scene spaces to 'speak up or do things' beyond making money. In this way their power to operate as 'the gay community' was also questioned:

> Yvonne: The majority of commercial LGBT stuff is focused around bars, clubs, pub nights, sex shops, that kind of thing. I think the community groups and the voluntary sector do provide more diverse choice and opportunity for different groups and different people. [But] I don't know if you could say that everybody was catered for. I'm not sure you can dictate to the commercial scene what it should or shouldn't offer. At the end of the day, commercial is commercial, and they're there to make money.

Claiming that Brighton's gay businesses were more than paid spaces of entertainment recreates an LGBT politics that gives businesses meaning and power in the reconstitution of LGBT lives. By denying this and the positioning of gay businesses as 'community', Yvonne challenged the necessary association of LGBT communities with the commercial, profitable and marketable. In this way she sought a different form of politics, one that contests gay bars as 'the gay community' and recognises their limitations in catering for and representing the diversity of LGBT collectives.

Conclusion

Scene spaces were constructed through a normalising of particular gay (*sic*) identities, bodies and practices. Where normalisation is in part achieved *through* acceptances in gay spaces, homonormativities can be more than replications of heteronormative structures. Instead they offer belonging for some who have been marginalised because of their gender and sexual identities. This is not to deny the privileges of those who felt included in the scene, but it is to point to the dangers of *only* recognising these privileges without engaging in ongoing exclusions that can also mark lives defined as 'homonormative' (see Brown 2008; Noble 2012).

However, not all gender and sexual dissidents are rendered ordinary even in supposedly accepting gay scene spaces. Exclusions from the scene are particularly painful because this can mean being multiply marginalised, both from mainstream and LGBT spaces. As an assemblage of interrelationships, practices and experiences the illusion of the scene as 'community', combined with the enjoyment of the scene made marginalisation from scene spaces more than just 'missing a good night out'. Not being part of the scene also related to not being part of the 'gay city' and the promise it offered. Yet a range of LGBT people felt that they should be able to access the ordinariness that scene spaces, and thus gay Brighton, was supposed to offer.

Although discussions of multiple marginalisation may suggest that the scene is either inclusionary or exclusionary, in this research we found that for some people the gay scene can be both. In CMIT people spoke about feeling like they belonged to the gay scene and gay Brighton, while simultaneously referring to the exclusions that they felt. Thus, scene spaces had multiple emotional attributes (positive, negative and ambivalent) with 'powerful psychological and ideological meanings' (Pritchard et al. 2002: 107). 'Scene bashing' through discussions that only attend to the difficulties and exclusions of gay scenes may neglect the importance of these spaces for many people, as well as how LGBT people at times found and created spaces for themselves that were welcoming. Thus ,there is a need to further explore the contradictory positionalities and attachments that were associated with these spaces.

Alongside discussions of marginalisations on the basis of identities, we also identified the importance of practice in creating inclusions and exclusions on the gay scene. In particular, there were social consequences to 'not drinking'. Further work (both in terms of support services and academic analyses) is needed to consider the place and importance of alcohol (and drugs) not only in creating scene spaces, but also as an aspect of social wellbeing and isolation. Beginning from an appreciation of the positive benefits experienced with drinking, and exploring what these spaces can offer in terms of validation, identities and safety, such discussions can avoid a re-medicalising (re-pathologising) of LGBT bodies.

The scene was important because Brighton requires a vibrant scene in order to *be* gay. Although 'the scene' in Brighton lacked cohesiveness, geographical imaginings of these places enabled them to be deployed as 'the gay community' for particular (often business or political) ends. In contesting particular power relations in the city, some sought to position Brighton's scene as purely commercial with no role in LGBT service provision and community development. Nonetheless in Brighton, gay men who control the scene still spoke for and as 'the gay community'. They held not just the monetary capital but also the cultural and social capital to take on the role of providing commercial scene spaces (see also Pritchard et al. 2002, Taylor 2007b) and defining LGBT culture. Chapter 8 further explores the gendered power relations that were created in and through Pride in Brighton and Hove, delving into the constitution of 'gay Brighton' through the deployment of 'gay community' and its association with particular groups and individuals who act as representatives. In the next chapter we further explore the multiple marginalisations that reconstituted LGBT lives in the whole city of Brighton, through examining bi and trans experiences of public sector mainstream services and LGBT community and voluntary services.

Chapter 5
Bi People and Trans People Under our Umbrella?: Contesting and Recreating Ordinariness

Jody: For example [name of gay male editor of media outlet] sent out an email recently saying 'Join the debate. The politics of Pride. What does gay community mean to you?' And I just wrote back a one-line email saying, 'well I guess I won't be there since I'm not fucking gay'. There were loads of experiences really where I just sat in room after room hearing about 'gay this, gay that, gay the other', from people who should know better, from people who are talking about the LGBT community. They occasionally correct themselves and say LGBT, you know that they're actually not thinking about bi or trans issues at all and they're not thinking about the differences between gay and lesbian issues at all. I'm tired of reading Stonewall reports that marginalise and exclude me. I'm tired of reading on the front that it says lesbian, gay and bisexual and then it's gay, gay, gay all the way through. I am tired of it. It's not a debate anymore. We're part of the community and that doesn't mean that our experiences are the same.

Persia (separate interview): There's still a lot of anti-trans feeling in the lesbian and gay world [laughs]. Massive amounts and an unwillingness to be associated with trans people, because it's like, well, we've made a nice neat little house here, where we're acceptable. We don't need a gang of weirdoes to come in and get people to associate us with you and then blow our whole [acceptance]. I know [trans project in Brighton] is now a safe place. Great. At least there's somewhere.

Introduction

Bi people and trans people were nominally part of the equalities 'gains' of the early twenty-first century. Yet, Persia and Jody speak to the exclusions and otherings that trans people and bi people experienced in this era. Using collectivities such as LGBT, differences in bi and trans identities and experiences can be elided and erased in favour of the dominant perceptions of the category, associated with lesbians and gay men (see Monro 2005, Richardson and Monro 2011, also Appendix 2). As has been well demonstrated, LGBT organisations, groups and individuals can present themselves as enacting bi and trans inclusion, while in reality perpetuating bi erasure, invisibilisation and trans exclusions (see for example, Angelides 2006, Barker and Yockney 2004, Browne and Lim 2010, Doan 2009, Hines 2007, Jorm et al. 2002, Oxley and Lucius 2000, Richardson and Monro 2012). Here, we begin from the premise that, although all LGBT people were ostensibly included both in terms of legislative change and as part of 'gay Brighton', 'they're not thinking about bi and trans issues at all' (Jody above).

We address 'they' in this chapter by examining public sector services and LGBT specific services. Moving beyond commercial gay ghettos that have been examined elsewhere and found to be exclusionary (see Chapter 4 and, for example, Doan 2010, Nash 2010, Hines 2010), we explicitly answer the call to examine whole cities in terms of sexual and gender difference (Brown 2008). Turn of the century inclusions of LGBT people and requirements to address social exclusion shaped a landscape that differed from previous eras. Prior to this public sector services and local authorities had used heteronormative framings that enforced normative assumptions regarding both sexualities and gender identities (see Cooper and Monro 2003, Cooper 1994, Monro 2005). The social inclusion agendas of public services were demonstrated in the move to the acronym LGBT as it became more widely used in the early twenty-first century. However, when required to cater for LGBT populations, local authorities predominantly recognised lesbians and gay men, side-lining bi people and trans people (Richardson and Monro 2011, 2012) and subsuming differences between L, G, B and T under binary sexual and gender differences (Barker 2012, Jorm et al. 2002, Monro 2005).

The violences of reducing LGBT to lesbian and gay differs from a positioning of all LGBT people as 'weirdoes' segregated from heteronormative worlds. Instead, recognising lesbian and gay men and conflating this with LGBT can produce marginalisations and effects that differentiate both trans people and bi people from lesbians and gay men. By putting bi lives and trans lives side by side in conversation, this chapter investigates the cross-cutting exclusions of being simultaneously within *and* outside of LGBT communities, exploring how some bi people and trans people failed to be part of the gay city.[1] This dual positioning draws

---

1   However this commonality does not imply that these identities can be conflated. While there can be little doubt that there are significant differences between bi lives and trans lives, these are addressed in this chapter due to the overlaps of exclusions of B and T from supposedly LGBT collectives. In our research, people in these groupings were often significantly more marginalised than lesbians and gay/non-trans respondents on a range of statistical measures (see Browne and Lim 2008a, 2008c). Moreover, these were the categories that were most noted by participants, service providers and others as needing attention. This speaks to the tensions of 'inclusion' of named categories. While it could be claimed that, to an extent, gay men and lesbians were 'fine' in the gay capital, this was noted to be insufficient in addressing the diversity of LGBT. The importance of using this category to ensure bi and trans voices are 'heard' will be addressed below, here we want to note the silences and erasures of other social differences. In particular, while there can be little doubt that raced, classed, abled, aged and other social differences were apparent in 'gay Brighton', the equalities legislation on sexual and gender difference and the alliance of LGBT forced attention to those who are named, yet marginalised, within these categories.

See also Appendix 2 for a full discussion of LGBT identities and our use of this category, as well as an examination of some of the critiques of these identities and the politics built from using them.

on and develops on understandings of multiple marginalisations (see Appendix 2). Multiple marginalisations enable us to explore the perspective of those who simultaneously 'won' in terms of sexual and gender legislation, but continued to experience exclusions on the basis of their gender and sexual identities.

We begin this chapter by examining experiences of mainstream services for bi people and trans people. Focusing on the materialities of basic needs, we delve into heteronormative experiences of marginalisation from the gay city. In examining mainstream services, we contend not only that heteronormativities continued to be salient considerations in equalities landscapes, but that these also operated differently for bi people and for trans people. We then turn to services that specifically target LGBT people and communities. Crossing gender (male/female) and sexual (gay/straight) binaries troubled the implementation of 'LGBT' services that often relied on seeing LGBT communities through the needs of cisgendered gay men and lesbians. While it is common to only articulate the problems and adversities faced by bi people and trans people, we seek to move beyond such hopelessness. Persia (above) points to the importance of having 'somewhere' to feel safe. We therefore finish this chapter by highlighting the productive value of acknowledging the 'othering' of trans people and bi people. Building on the possibilities of gay imaginings of cities developed in Chapter 3, we contend that this was a result of the *expectations* of how equalities landscapes *should* be enacted in this city.

Heteronormativities: Material Implications for Identities beyond Sexual and Gender Binaries

There was an expectation in Brighton that bi and trans differences would be recognised within, and catered for by, mainstream public sector services, given presumptions about the size of 'the LGBT community' (see Chapter 1) and the idealisation of the gay capital (see Chapter 3). In highlighting exclusions from these services through the lens of multiple marginalisations, this section explores the materialities of heteronormativities that were recounted by trans people and bi people. In this way it illustrates that gender and sexual identities should be considered in terms of inequalities beyond discussions of the fracturing of sexual and gender categories (see Appendix 2 for a discussion of these debates and McDermott 2011).

While reliance on public services varies by social differences, such as class, a greater proportion of trans and bi people in the CMIT sample were reliant on state provided or publically funded services, compared to others in the research. More than this, those in this grouping were more uncomfortable using these services. In contrast to the expectations of the gay capital, in the Count Me In Too research more bisexual people (26 per cent, n. 11) reported feeling uncomfortable because of their sexuality or gender identity when using mainstream services than gay men (12 per cent n. 49, p<.002, see Browne and Lim 2008a). Similarly, trans people (36 per cent, n. 15) were more likely than non-trans people (16 per cent, n. 115) to feel uncomfortable because of their sexual or gender identity when using mainstream services (p<0.05, see Browne and Lim 2008c). While these figures point to differences in bi lives, trans lives and those of gay men and non-trans people, the majority of both sample groupings say that they do feel comfortable using key services.

In the qualitative data, trans people and bi people painted a mixed picture of their experiences of using public services. These varied both between people and between services accessed by one person:

> Sashi: I have to have yearly medicals because I'm on incapacity benefits and it's the only environment where I'm really keen not to mention it [bisexuality], because I'm really not sure it'll be received caringly or carefully. But it's quite a spiky environment anyway. It can be a very kind of combative thing, to have a medical, because essentially you're kind of being checked out to see if you're faking. So I'm really conscious that I never talk about my sexuality there ever, because I just wouldn't want them to know more about me than they need to know.

> Lucy: I did engage with a lot of the government services in the last months. I've engaged with the mental health community and I had to fill in a form and I was quite clear I was bisexual and transgendered. The bisexual statement actually got an odd response, the transgender one seemed to be okay, but then it's a kind of freezing with the lady that I was speaking to. I applied to go on the council housing list and I wrote a letter about my history and that was kind of interesting because I went into the fast track place and there was a kind of shock. The guy at the counter [it] was quite clear, he [was] gay and he was trying to be helpful, but it was just like there was this shock, a sort of recoil to what was I was presenting them. I've been quite straight forward with housing benefit people as well, because I felt it was important. So I think I just get mixed response coming out. (Bi focus group)

The erasure of bi identities has been identified as a key issue in use of services (Barker and Landridge 2008; Barker and Yockney 2004) and, in contrast, for trans people in this research, hyper-visibility was perceived to be a key danger. For Sashi, where there was a reliance on others judging your 'truthfulness' in order to receive particular services, bi identities can be more 'than they need to know'. Lucy in contrast 'was quite clear' about her bi and trans identities, pointing to the need to examine the experiences of those who identify as both bi and trans, and the intersections of these identities that are not always neatly delineated. In spaces that were perceived to be 'dangerous', bi identities can be carefully negotiated yet, as Lucy notes, the intersections of bi and trans identities can be met with multifaceted responses. Her narrative shows how an individual can experience the state as diversely enacted (see Chapter 6), with sexual and gender identities taken into consideration by some public sector workers (and 'fast tracked') or policed by others ('a sort of recoil'). These narratives together point to the heteronormativities (as well as homonormativities) that continued to pervade the gay capital and how they affected bi people and trans people.

As Lucy identifies, housing is a key way in which trans people and bi people come into contact with state provided services. In Chapter 3 we discussed how some LGBT people who migrated to Brighton did not easily secure safe housing and this was one manifestation of how the gay capital failed to meet its expectations. Social housing was allocated by local government to those who were seen as having pressing housing needs using specific sets of criteria. Living in social housing points to issues with income and housing security and a reliance on the state. Similarly to the use of public sector services, almost a third of trans people who responded to the questionnaire lived in social housing (29 per cent, n. 12). Bi people were also more likely to live in social housing and less likely to own their own homes compared to lesbians and gay men.

It was not only the likelihood of living in social housing that was differentiated by sexual and gender difference. Finding accommodation was also problematic, even where this was not reliant on state support. Trans people were almost twice as likely to report that they struggled to get accommodation compared to non-trans LGB people (see Browne and Davis 2008). National and local research demonstrates that it is common for trans people to have difficulty finding safe, suitable accommodation, particularly during transition (see Whittle et al. 2007, Cull et al. 2006). Transphobia was reported to be an issue when seeking accommodation in the private rented market after transition. This illustrates the ongoing experiences of marginalisations that need to be negotiated:

> I couldn't view the accommodation. I got my partner to do that because landlords discriminate against trans people. (Questionnaire 212)

This response points to how prejudice manifests itself, creating housing vulnerabilities for some trans people. It is important to note that the perception of 'discrimination' as important as its manifestation in actions from landlords (see also Browne 2006).

It was not only in finding accommodation that trans people and bi people experienced prejudice. Discrimination from private rental landlords adds to evidence of experiences of domestic abuse from partners, children and families (see Whittle et al. 2007). In the CMIT research trans people were more likely to experience domestic violence and abuse when compared to non-trans respondents (see Browne 2007b). Finding safe housing may not be straightforward. There was evidence of ill treatment within social housing that was supposed to offer safety and accountability:

> When I was in [a] council B and B and complained about transphobic victimisation by hotel owners, the council sided with them. (Questionnaire 828)

Discussions of gay/lesbian home-spaces (Valentine and Johnstone 1995; Gorman-Murray 2007, 2008) can overlook the importance of the materialities of housing. In this research, the basic needs of trans people and bi people sometimes went unfulfilled, in part because of their sexual and gender identities. This indicates a need to examine sexual and gender difference in ways that account for the material effects of difference (see also Taylor 2007, 2009).

Participants in the bi focus group discussed the lack of affordable housing as a key issue of living in the city. Celebration of rising house prices in Brighton and nationally during the period up to and of the research (see Chapter 3) was met with disdain by those in the bi focus group, many of whom struggled to afford to find housing in the city. Perhaps unsurprisingly then 33 per cent (n.15) of bi people in this research had experienced homelessness in their lifetimes (16 per cent, n. 7 in Brighton), compared to 22 per cent of lesbians (n. 49, 8 per cent in Brighton, n. 23) and 19 per cent of gay men (n. 80, 12 per cent in Brighton, n. 53) (see Browne and Davies 2008, Browne and Lim 2008a). Similarly, over a third of trans people in this research had experienced homelessness at some point in their lives (Browne and Lim 2008c):

Persia: I was homeless down here. I had some real trouble in this town. Mid '90s and there I was and I was still in the transition phase and my work disappeared. I had nowhere to live.

I was sofa surfing.[2] This is what they call the hidden homeless. I was never on the streets. On the surface I looked fine, for all the problems. I fell down the big black hole.

It was the Brighton Housing Trust [that] sorted me out with a place to live. [I had] a huge concern about how people would react to you and a great deal of fear. The Brighton Housing Trust handled that for me. So there's the evidence of having a social institution, somebody looking after people. [They] looked after me and that made all the difference in the world. I was really lucky.

[When] I went to the Brighton Housing Trust, I had to wait outside to get in. I was mortified. [It] did me the world of good, you know, that middle class pride, that was that out of the way. In the end I went in, in a little room, and the guy said 'What can I do for you?'. I said 'I'm destitute, I'm homeless and I'm transsexual

---

2   Being homeless in Brighton did not necessarily pertain to living on 'the streets', and the geographical imagining of homelessness in public spaces was queried in our data. We found that LGBT people relied on friends, sofa-surfing and sex work in order to find accommodation. This data points to the ways in which bi and trans people can be particularly affected by homelessness and our data pointed to a link between experiences of domestic violence and abuse and homelessness that had particular pertinence for bi and trans people (see Browne 2007b, Browne and Lim 2008a, 2008c).

and I really need somewhere to live'. He said, 'okay hold on a second' and he came back in about five minutes and said 'right, we've got a good place for you'. So there was not a trace of anything. It was just something to be handled by him and that made all the difference in the world, just to have people that are open and *the council has been learning recently about these specific issues.*

I had a GP who was just completely totally supportive in all respects, totally respectful. And then I had the Brighton Housing Trust. So that set the foundation and then from that I built my strength up again. (Emphasis added)

This account of a positive experience of housing services in Brighton reiterates the importance of housing and supportive health care in 'setting the foundation' that enabled Persia to use services to 'build up her strength again'. This demonstrates that the materialities of housing are key to engaging with the wellbeing of bi lives and trans lives (as well as contesting class privileges that some may have before transition). Persia's narrative indicates that, more than mental health and other healthcare services, housing services were vital in providing support for her at vulnerable period in her life, around her 'transition phase' (see Browne and Lim 2010, Browne and Bakshi 2011, Lim and Browne 2009). The multiple points of contact with state services, to health, housing, financial support and so on, created a 'huge deal of concern', but state actors *also* had the potential to make 'all the difference in the world'.

Reports of positive experiences were associated with key people providing services in individuals' lives (such as GPs). This reiterates the importance of individual actors in contesting (Cooper 1994) as well as potentially enacting heteronormative exclusions and othering. Not only were place imaginings then informed by these experiences, but Persia also mentions Brighton and Hove City Council as 'learning' from these experiences. Augmenting this with expectations of what *should* happen in Brighton as discussed in Chapter 3, LGBT activists understood 'the gay capital' as working towards its unattainable ideal of offering all LGBT people ordinary lives. For trans people and bi people there was a failure to live up to these ideals, including not meeting basic needs. This lead to disappointment and hurt, while those individuals and services who did cater for bi people and trans people 'made all the difference in the world' and reiterated the 'difference' of Brighton in comparison with other places.

LG(BT) Services?

Persia noted in the quote that started this chapter that the margins of LGBT identities can be recreated as 'freaks' in order to normalise certain cisgendered gay identities. It is increasingly understood that bi people and trans people can experience discrimination, prejudice and abuse from (homonormative) gay and lesbian spaces, contesting the view that these are LG*BT* friendly (Doan 2009,

2010, Hines 2007b, 2010, Monro 2005, Nash 2010). While there has been some discussion of bi and trans exclusions from LGBT scenes and spaces (see, for example, Doan 2007, Hines 2007b, 2010, Monro 2005, Nash 2010, 2011), LGBT services have yet to be fully scrutinised in this way. Here we examine bi people's experiences of LGBT services in Brighton, which aimed to provide support for fundamental life issues and which were supposedly actively seeking to engage the spectrum of LGBT people. The focus on bi people's perceptions of LGBT services allows us to examine the importance of straight/gay binaries in the design of LGBT services.

Bi respondents were less likely than lesbians or gay men to prefer using LGBT specific services (see Browne and Lim 2008a). Some bi respondents felt that even services offered by specialist LGBT organisations were potentially damaging to them due to failures to respond helpfully to their bi identity:

> Chris: If you go to a lesbian and gay service and you experience biphobia in the place where you're trying to be your most open and you're at your most vulnerable, that's really dangerous. It really does wreck people's lives and mental health services need to a) be more educated about bisexuals; and b) to just have people that you can go and see, especially LGBT services, because they're already setting up for this niche thing. (Bi focus group)

Setting up dedicated LGBT provision does not necessarily achieve inclusion across the diversity of this category. Not all experience these spaces as safe. This has material effects such that, while the category of bi might be contested, fluid and multiple (see Appendix 2), biphobia is keenly felt and can be 'dangerous' beyond experiences of physical violence from heterosexual others (see Chapter 7). Chris points to the betweenness of being both within the 'niche' of LGBT and yet simultaneously not catered for by this 'specialist' provision.

Zara identifies two aspects of her identity, bisexuality and ethnicity, where she fears specialist provision may fail to respond appropriately. Her intersectional positioning is manifest in her seeing services as 'requiring testing' or even being unusable, where they may not be able to cater for her needs:

> Zara: I lived in Brighton for a couple of years, with a monogamous male partner. It's incredibly hard to walk into an LGBT service when you're a girl going out with a boy and feel like you belong there. It's incredibly hard to do that. So I didn't go anywhere near anything LGBT.
>
> Kriti: Do you feel confident it would be appropriate for you?
>
> Zara: No, actually, both on the grounds of bisexuality and of ethnicity stuff, and there is a bit of me that probably should. I should go and test this out but you ask and my instinctive response is just no.

Kriti: My instinctive response is *not* to go and test it out [laughter] because if I need support, you don't wanna go in there, that's the worst possible time to test it, is when you need support. (BME focus group 2)

Bi people can contest the binary understandings of gender identities and sexualities on which lesbian and gay services are built. As such, bi identities can question the premise of a lesbian/gay service that relies on assumptions of 'same sex' attraction. Not only might feelings of belonging be contested by Zara using LGBT services with a monogamous male partner, her intersectional identities of relating to ethnicity and sexuality furthered her negative 'instinctive response' (see Catungal 2013; Kuntsman and Miyake 2008). The assumption that services for (white) lesbians and gay men can be opened up to include bi people on the basis that public services have expertise in dealing with issues associated with same sex relationships is flawed. This is potentially damaging to bi (and trans) people who become 'included' when the acronym LGBT is adopted. The spaces of LGBT services were at times effectively lesbian and gay, so bi people (and trans people) felt disqualified and excluded from such spaces. Those experiencing mental health difficulties in particular may not want to take the personal and emotional risks associated with 'testing' their belonging in such spaces, even in the gay capital where these services were supposed to be better.

In addressing the material consequences of a lack of understanding of sexual lives beyond gay or straight, some asserted that training was required so that service providers would 'know how to start the conversation, know how to treat us'. Brighton, as an imagined place, was active in forming expectations that LGBT services could become inclusive of the spectrum of LGBT identities. It was expected/known that there were people in the city 'who have a lot of experience training people in this kind of things, or who run these services' (Chanda, bi focus group). Such calls for training locates the issues that bi people and trans people encounter in using services with individuals who deliver such services. It suggests that there are experts who are outside of these organisations who can be brought in to transform the services provided for LGBT people. While individuals can make a positive difference through their actions and attitudes (as described above), this approach fails to address the systemic failures in catering for bi people through the use of a bifurcated model of sexual difference (gay/straight).

Knowing 'how to treat us' then was not a straightforward proposition. The complexities of bi identities, which include intersections such as ethnicity and are key to academic writing (see for example Taylor et al., 2011), have material manifestations including, as we have seen, housing and LGBT services. These identities needed to be understood in multifaceted ways, requiring investment by services, as well as bi people:

Rachel: I think for me it's central to have the willing to challenge biphobia, because it's very difficult to do within the community that is supposed to be our community, it's so hard [that] it's painful. I think that work's got to be done.

Audience member: A lot of bi people experience a great deal of biphobia in lesbian and gay space, so I guess it's a kind of a question about how bi-safe space can be brokered, won and kind of made part and parcel of LGBT [name] services and in a wider sense for everyone else as well, about how it can be made part of LGBT space. If it's supposed to [be a] LGBT space.

Rachel: I agree with that and I think that there are lots and lots of complexities within. I can sit here and go safe space, safe space, but actually, yes, there are loads and loads of subtle nuances around that. What is [a] safe space to one person is not a safe space to another person, based on a whole complex set of rules and reasons. I don't really have all the answers to that yet, but I think a part of what [a specific post] is going to do is look at that and speak to people like yourself to find out a bit more about what that means. (Consultation meeting, April, 2009)

Considering the complexities of safe space and recognising that these require nuanced attention and engagement with bi people has the potential to enable a move beyond simply recognising biphobia as a problem with little hope of addressing it. There was desire amongst some public and LGBT services to understand what it means to create 'bi-safe space' as an integral element of LGBT spaces. This recognises the betweenness of being simultaneously within and outside LGBT collectives and provisions. Biphobia can occur from within and bi people may need different provision to lesbians and gay men.

Recreating Ordinariness in the Alternative City

Discussions of provision that address the failings of the gay city are underpinned by acknowledgement of the failure, so far, to achieve LGBT inclusive spaces in Brighton. The momentum generated through imaginings of Brighton as a place where there is the possibility of making things happen because of failure is apparent in this 'so far'. Not only should Brighton be better but also, at the time the data was collected, the possibility existed that bi people and trans people in Brighton could be catered for by LGBT services. The desire for specific LGBT as well as inclusive mainstream provision was part of reconstituting the 'alternative' city. The hopelessness that can emerge from critical thinking (Sedgwick 2003) or 'cynical resignation' (Halberstam 2011, see Chapter 1) can also be questioned through listening to the possibilities of place articulated by some. In Brighton this provided possibilities for bi people and trans people to create moments and spaces of ordinariness in the city.

Gay Brighton promised much, not only to lesbians and gay men but also to bi and trans people. Bi people and trans people, along with lesbians and gay men, spoke of 'running' to Brighton for the safety and acceptance that the city supposedly offered:

Bobbie: I sort of ran away [to] Brighton. It was about four or five years ago, haven't regretted it.

Ramsey: The city itself, I've not personally found any problems with it. I've found it quite kind of nice and accepting and it seems to be a city that prides itself on being weird and wonderful and I think there are probably quite a lot of people who keep quiet about their own dislikes or uncomfortable areas because they like the fact that's a weird and wonderful city and I think that I've always found that it's quite a nice, generally a nice place for everyone.

Ruth: Generally Brighton feels a lot safer than I have done anywhere else. I do get funny comments walking up St James's Street but I just let it wash over me, I don't take any notice. But I live on the street, so it's like I'm part of the community and I know people and I recognise people in shops and things like that and people I can chat to and always bump into friends down there, so it never really bothers me. (Bi focus group)

The 'alternative city' of Brighton was seen as providing space for difference and enabling safety (or at least, not feeling 'unsafe') for some bi people and some trans people. Reflecting the themes in Chapter 3, the city was understood as facilitating 'weirdness' in ways that leads people to 'keep quiet about their own dislikes'. The comparison with London (and other places where participants had lived) was important in imagining Brighton as different to 'anywhere else I've lived'. While people continued to 'get funny comments', bi people and trans people found safe and accepting spaces, and thus claimed a stake in the ordinariness of the extra-ordinary city. Not only did Brighton create possibilities for experiencing this ordinariness, it also was created through it as an extra-ordinary city where bi people and trans people were (ostensibly) included.

Where mainstream public service provision sought to speak with trans people and bi people, this went some way to address the experiences of discrimination, exclusions and fears identified above:

Geraldine: A lot of those people in the [name of trans support group] were really scared about the police and the way they'd react to them, and taking officers in uniform to the group and just chatting to people, it kind of broke down that barrier. That made a huge difference and that showed them that the Police were wanting to change and to assist and then also the trans employment policy as well, to actually try and recruit trans people. I was really surprised by how supportive the police were. They took it all on board fully and stuff that I thought probably might not actually go through, did go through and they were all supportive of it.

In Brighton, some within certain mainstream services (such as the police) began to recognise and act on the legal (see Chapter 7) imperative to cater for LGBT people.

This was not only about addressing lesbians and gay men, some also understood that trans people and bi people have different requirements for safe spaces. Recognition of the differences within this category was manifest in seeking to include diverse LGBT people in meaningful dialogue in spaces where they felt safe. This had the potential to make them ordinary and part of the extra-ordinary city.

Such ordinariness was not necessarily about subsuming these differences into a hetero/homonormativity but, in certain contexts, led to acknowledging and addressing the concerns of bi lives and trans lives on their terms. For example, Brighton City Council Equalities Forum was scrutinised about bi issues and trans issues (before this forum was removed by the Conservative administration in the mid part of the decade). Through this critique, equalities initiatives that solely addressed lesbian and gay identities were highlighted as inadequate. In this way, activists were able to force local government services to address the diversities of L, G, B and T. Naming bi and trans in this way questioned the homogeneity of LGBT and did enable (limited and contingent) forms of bi and trans inclusion, visibility and recognition in gay Brighton in the first decade of the twenty-first century. This lead to the formation of a bi and trans working group which spoke with service providers, and others, addressing key needs identified by representatives of these groupings. Thus, geographical imaginings of place that saw Brighton as LGBT friendly served to facilitate work that addressed specific identities and differences within this collective. Where LGBT work was seen as key to 'progress' in the city, the political imperatives of failing to meet the ideal of gay Brighton meant that mainstream services were encouraged to engage with the heterogeneity of L, G, B and T. Richardson and Monro (2010) argue that the need to label people within particular groupings in order to cater for these groups reproduced illusions of the homogeneity of bi and trans within local authorities. Our research suggests activisms both deployed and questioned labels such as LGBT (see Chapter 6 and also Browne and Lim 2010, Voss et al. forthcoming). Having written about some of the complexities of trans activisms and bi activisms elsewhere (see Browne and Lim 2010, Voss et al. forthcoming), we now focus on how cis-gendered/non-trans people, lesbians and gay men sought to work with and for both trans people and bi people. In this way we continue to question the hopelessness that discussions of marginalisation can engender and the binaries of 'cynical resignation' or 'naïve optimism' (Halberstam 2011).

Although differences and nuances were at times elided, key informant interviews evidenced a complex understanding amongst some lesbians and gay men of trans issues and bi issues. For example, lesbian and gay groups in Brighton took steps to ensure that they were 'truly' inclusive of the B and T of LGBT. This included seeking to train workers so that they were better able to support bi people and trans people, and engaging with specificities associated with the range of identities included in the LGBT grouping, before adopting the term 'LGBT' to label their service. Following this a range of follow-up assessments were actioned to address the ongoing difficulties bi people and trans people discussed:

Natalie: We've been working at [name of organisation] recently [2009] on a bi project to really assess how inclusive we are about people. We've used a kind of focus group that's made up of members of the bi community who have been working with me on putting in place a strategy for making sure we're as inclusive as we can be.

Recognising and responding to a need to 'assess how inclusive we are' shows that there was awareness in Brighton that LGBT services were not attaining the ideals of LG*BT* inclusion. Some services were very aware of the need to work with bi people (and trans people) 'making sure we are inclusive as we can be'.

The dissonance between the imaginings of Brighton (as 'better', 'inclusive', 'gay friendly') and the realities of Brighton life for trans and bi people, offered the potential to rework who could be ordinary within the banner LGBT. While some gay and lesbian key informants understood their own lives as commonplace in Brighton, they did not see everyone in the category of LGBT as part of this ordinariness. However, they believed that this should be the case. Rather than the 'freaks' that Persia describes in the opening quote, gay men and lesbians spoke of the 'horrific' issues facing trans people in particular, and how this was not appropriate in the gay capital of the UK:

Andrew: I think the trans issue, or the issues facing the trans community, are horrific. They've got the most difficult position in society in many ways. I mean the figures around prejudice, abuse and just trying to get on in life are just off the scale. I don't actually think the identification with LGB [is] doing that community any justice. I do think the T can get forgotten and I also think we're mixing things there. We're mixing gender identity and sexuality. And those things are not the same thing. That's the whole point of trans. If we can help by joining together, I'm happy to help, but sometimes I do question it. Trans is just a completely different agenda. We need to be dealing with the issues around the health service and their access to appropriate health, dealing with violence and all the stuff in terms of accessing service[s], the treatment they get on the streets. There's so much that needs to be done about that and it's colossal and really serious and we need to do more on it, is my view on that. I'm not trans.

Andrew saw a hierarchy of discrimination in Brighton, which places trans people in 'the most difficult position' and sees trans issues as a 'completely different agenda'. While questioning the coherence and usefulness of T in the context of LGBT, Andrew is committed to working with/as an LGBT grouping 'if we can help by joining together'. As he was strategically placed as a senior figure in a public sector organisation, his awareness of trans issues points to the possibilities that existed for (perhaps limited) 'inclusion' and the use of allies in addressing the exclusions of trans people. This is not to suggest that these recognitions 'worked' positively for trans people, nor does Andrew indicate understanding of the complexities of trans communities. Rather, this points to the possibilities of LGBT politics in Brighton.

Brighton, a small city with a metropolitan gay reputation, placed bi people and trans people simultaneously within and yet separate from LGBT groupings. This enabled some trans people and bi people to 'ride the wave' of LGBT inclusion. There was a variation in the geography of separating L, G and B politics from T politics. The separation of 'us', usually gay men and lesbians, from 'them', trans people in this case, was reportedly welcomed more readily by groups based in London, compared to activisms in Brighton. In the latter, the most vocal trans activists (and bi activists) understood themselves as part of a bigger LGBT grouping that could benefit from lobbying for change on the basis of the numbers of LGBT people in the city (see also Browne and Lim 2010). They believed that they have benefited from a historical recognition of the numbers of lesbian and gay people in Brighton. This implies that, because LGBT people were to some degree accepted and included within the culture of Brighton, trans people and bi people benefitted from some forms of acceptance and inclusion as a consequence of being grouped along with lesbians and gay men. More than this, it indicates the need to address the specifics of place when examining the possibilities and limitations of LGBT politics that is inclusive of a broad spectrum of bi and trans people. Being part of the gay capital gave some bi and trans activists traction to live life in ways that contested their marginalisations, this may not be the case elsewhere, and indeed being subsumed under the banner LGBT might result in further issues of exclusion and erasure.

There was also the possibility of creating trans spaces and bi spaces. For example, deploying trans identities enabled some trans people and allies to support each other in creating safe space. This was for some 'a sacred place':

> Jill: [When] the [name of group] started it was a monthly get together but, as that started to die, I thought, 'Ooh, what would be great would be a drop-in because then some people have got somewhere to go'. So we started a drop-in and it's going and it's flourishing and they're all coming and it's great and I love it and what it is, it's a sacred place. You go into that room, you're safe. No matter what you are, no matter where you come from, no matter how you look, but you come into that place, no matter what you're welcome and you're fine as you are and just to find that place in the world is a rarity, believe me.

> I'm all right. I'm cool I get away with it, but I didn't used to and that was awful in those days. So I think it's a wonderful [thing] that the [name of group] exists and it's been taken forward by the great people and it saves lives. (Consultation event, April, 2009)

Supportive spaces for 'what you are', 'where you come from' and 'how you look' were said to be a rarity for trans people even in the gay capital. While fraught with tension at times, they were valued by many and created alternative conditions of ordinariness in specific spaces and through encounters that challenged cis-gendered norms.

Similarly the imaginings of what Brighton 'should be' created bi-specific spaces and a sense of bi communities. Participants in this research both created and used spaces where they could 'finally feel like [they were] at home'. Similar to Hemmings (2002), bi participants described temporal spaces, which realised the 'fantasy of discrete bisexual, identity and community' and talked of the ways this 'came true' at bi conferences and events, such as BiFest (Voss et al. forthcoming). In the social and support group, Brighton Bothways, bisexual 'desire, identity and community formed a core around which "other" identities are negotiated' (Hemmings 2002: 145). In discussions of BiCon, a national bisexual convention, this has been described as 'BiTopia' (Anderlini-D'Onofrio 2011). These temporary spaces were created by bi people (and trans people) themselves, forming their own spatial and temporal ordinariness.

New conditions of ordinariness made possible through the creation of bi and trans specific spaces were understood as questioning and challenging lesbian and gay hierarchies. This was undertaken without homogenising bi communities and trans communities or negating the heterogeneities of those grouped under bi and trans labels. This did not mean that there were no commonalities and there was evidence of productive linkages between bi activisms and trans activisms. Working together and for each other was premised on having considered gender and sexual identities in ways that others 'don't normally have to think about':

> Stevie: I think probably the most inspiring contact and work I've ever done within LGBT communities has been with trans people. So much more inspiration from the trans community than I ever have from the lesbian and gay community. And so much more interesting discourse as well around gender fluidity and sexual identity and all this kind of stuff, because trans people just have this unique experience of taking hormones. I mean where does gender live? Where does sexual identity live? Where does it all live? Does it live in your body which is changing? Does it live in your hormonal identity which you're administering to yourself or having administered to you? Is it in people's ideas of you? Is it in how you're perceived? Do you know who you are when people name you in the street? Do you know who you are when you pass, when you fail? Those questions are so alive for trans people. They're pioneers. Trans people are pioneers of gender and sexual identities in ways that we don't normally have to think about. [I have] experienced some of that pressure coming out as bi in community, but to have to live with that pressure constantly. My absolute heroes, every trans person. As a person who doesn't identify as trans, I think it's really important for me to be involved in trans work and support trans people.

Conclusion

In Brighton during the first decade of the twenty-first century, there were possibilities in working as a collective LGBT grouping that sought to address the exclusions and marginalisations of bi people and trans people. They were seen as having *yet* to achieve ordinariness and in this failure there were possibilities. However, this was a complex and nuanced proposition. Elsewhere we have shown the complexities of this 'inclusion' and the potentials and problems it affords (see Lim and Browne 2009). Here it is important to note that some trans respondents voiced concerns about inclusion of trans under the banner of 'LGBT', saying that this sometimes entailed a certain level of invisibility, and bi participants spoke extensively of bi erasure within this collective (see Browne and Lim 2010, Jody above; see also Barker and Yockney 2004, Jorm et al. 2002).

There can be little doubt that there were pitfalls, imperfections and failings in coalitions between bi people and trans people, as well as amongst the broader LGBT collective. As we have seen, trans people and bi people suffered painful exclusions from other LG(BT) people as well as, at times, from each other. However, there is a danger in reading trans lives and bi lives only through suffering and exclusions. In Brighton there was an active decision by some individuals and groups to acknowledge bi and trans identities as different but part of the 'we' under the umbrella 'LGBT'. LGBT people sought to work with, and respected, differences, complexities and fluidities, producing new normalities that were expected of the gay capital, but did not necessarily materialise throughout mainstream LGBT spaces.

The material effects of multiple marginalisations, when accessing and using both mainstream services and LGBT specific services, point once again to the importance of investigating how bi people and trans people live and experience their lives beyond theoretical inferences (see for example Namaste 2000, Noble 2006, Prosser 1998, Wilkins 1997). There is a pressing need for work that looks not only at the complexities of trans and bi identities, but also to how this creates material discrimination that affect basic needs, such as housing. This brings discussions of redistribution to the fore and calls on scholars to examine sexual and gender identities as key facets in the distribution of key public services.

LGBT inclusiveness asks for a form of political empowerment that questions the boundaries of straight/gay, male/female. However, gay and lesbian inclusions can be based on these binaries. This has been questioned through examinations of exclusions from gay and lesbian/LGBT spaces (see for example, Doan 2007, Hines 2007, Nash 2010). Through enlarging this focus to include mainstream public service provision, we illustrated that not only did heteronormativities continue to be salient considerations, but also that these operate differently for bi and trans people, compared to lesbians and gay men. This indicates a continuing need to address heteronormativities in ways that can be overlooked when there is a focus on homonormativities or a celebration of the world 'we' (often lesbians and gay men) have won. Thus, alongside examinations of homonormativities and exclusions from civil rights gains, explorations of ongoing heteronormativities that account for variations between lesbians, gay men, bi people and trans people are necessary.

In this research, the expectation that this city should cater for all LGBT people also opened up possibilities. Bi activists and trans activists were able to claim a stake in the gay capital of the UK. New conditions of ordinariness were recreated in and through imaginings of what Brighton should become in light of its contemporary failings. Consequently, the processes, marginalisations and possibilities of bi and trans exclusions and inclusions are *created by* place. In other words, imaginings of place reconstitute bi lives and trans spaces, politics and lives in gay Brighton. In the next section of the book, we move to further examine the possibilities of activisms that sought to make LGBT people ordinary and included as part of Brighton. We continue our focus on the public sector, contributing to the call to move beyond 'gay quarters' to explore the whole city (Brown 2008). In the next chapter we begin by outlining the changes from politics that opposed state and oppressive legislations, towards activisms that worked within inclusive state apparatuses that were created in the first decade of the twenty-first century.

Chapter 6
Ordinary Activisms:
Beyond the Dichotomies of
Radicalism/Assimilation

Introduction

Although there are numerous examples of lobbying from within state structures to enable legislative gains for LGBT people over previous decades, the activisms which emerged in the UK at the beginning of the twenty-first century existed in very different contexts to those identified in the 1990s. Those of the 80s and 90s sought to change legislation or to work against heteronormativities in institutional spaces (Cooper 1994, 1995, Cooper and Monro 2003). By the late 1990s the position of acting from 'outside' of the state was coming into question where 'lesbian and gay [*sic*] work no longer represented opposition and challenge to hegemonic state forces' (Cooper 2007: 933–4). This challenges activisms that rely on resisting heteronormative state and regulatory processes to oppose. With the lack of opposition posed by lesbian and gay work, there has been a turn to contesting homonormativities created through sexual (and gendered) equalities legislation and cultural forms (see for example Brown 2007; Richardson 2005; Stychin 2006). Those who seek queer normativities can assert that the gains of civil rights and equalities lead only, and necessarily, to a loss of radicalism and a demobilisation of 'queers' (Sears 2005). We contest this view, recognising the productive value of the diversity of activisms. Rayside (1998) contends that 'engagement with the political mainstream has not led the gay and lesbian movement in uniformly less radical and more bureaucratic directions. It coexists with other strands of activism more wary of and antagonistic to the mainstream' (1998: 3, see also Cooper 1995, 2011, Weeks 2008).

In this second section of the book, *Ordinary Activisms*, we explore activisms that seek to make LGBT people ordinary and in-place, where once we were deviant and out of place. Such activisms we will show can be both, and at times simultaneously, cooperative and antagonistic. We investigate the work of LGBT activists who engaged with partnership work from within state organisations to explore activisms beyond binaries between them/us, enemy/ally. Rather than addressing engagements with the political mainstream as assimilations, which they sometimes are, and conformist, which they undoubtedly are, or merely selling out, we seek critical considerations of the dichotomy of radicalism/assimilation. In this way we extend our challenge to the binaries of 'cynical resignation' or 'naïve optimism' (Halberstam 2011: 1), exploring alternative activisms that are not necessarily outside/beyond state control.

We begin this chapter with poststructural conceptualisations of 'the state' and use this to unpack the radical/assimilationist binary that can underpin forms of sexual and gender politics. We do this in order to allow for the possibility that working with and from within public services can be considered activisms that create the possibilities of living differently, alongside the queering of state apparatuses. In the second section we examine the changing landscapes of LGBT political activism. This moved from politics that opposed overt and state supported heteronormativities in the 1990s, to the equalities gains of the early twenty-first century that created common sense validity to LGBT work. We use this backdrop to scrutinise the possibilities of partnership working where LGBT groups and people were part of citywide committees and were to some extent included in provision from policy makers and service providers. In the final section, we consider LGBT community and voluntary groups who work with public services, through the lens of Spectrum, the LGBT forum in Brighton. This offers insights into the multiple ways in which activisms formed through partnership working can deploy tactics of resistance *and* collusion, as well as the costs of working with 'the state'.

Conceptualising the Poststructural State, Querying Radical/Assimilationist Dichotomies

While there has been much discussion of the state in queer activisms that seek to disrupt normativities, far less attention has been paid to the implications of poststructural considerations of the state. Here we outline some of the ways in which the state can be seen as messy and performed, before moving to examine the implications that this has for considerations of the dichotomy of assimilation/radicalism. This dichotomy can be used to critique equalities gains, seeing some forms of activism not only as inferior but also as selling out. In this way the section outlines the theoretical underpinning of our discussion of ordinary activism in this section of the book, particularly as it is manifest through working with and within the state.

The state can be seen as part of our everyday lives, recreated through assemblages that are diversely and unevenly manifest and performed. When considered in this way, its effects or effectiveness can be viewed through how it is enacted, by exploring the diverse outcomes and possibilities of 'lines of operation' (Allen and Cochrane 2010, Painter 2006). Here, we follow Painter in contending that the state is constituted through mundane practices, such that 'everyday life is permeated by stateness in various guises' (2006: 753). This calls for a focus on the actualisation of the state, pushing us towards giving 'weight to the heterogeneity, complexity and contradictoriness of state institutions' (Painter 2006: 764). Thus, there is a need to examine the state through 'ordinary practices' that actualise and recreate daily life in unpredictable and unintended ways, such that in practical terms, the state may not be effective in 'achieving its regulatory goals' (Lloyd 2005: 452, Ahmed 2012, Clarke et al. 2008, Stenson 2008, Painter 2006).

The state is therefore heterogeneous in how it is performed, refuting the illusion of fixity, stability and coherence. Thus states operate (and are recreated) in geographically uneven and inequitable ways (see for example Ahmed 2012, Allen and Cochrane 2010, Butler 1990, Cooper 1994, 1995, Cooper and Monro 2003, Painter 2006). This can offer progressive potentials, but it can also mean that certain equalities (and what might be seen as socially progressive) legislations are not enacted across the legislative constituencies that they are supposed to encompass, setting up what Cooper terms 'defacto firewalls that impeded the ability of lesbian and gay politics to permeate many local authorities' (2006: 935, or as Ahmed, 2012 notes is discussed through the metaphor of brick walls).

Conceptualising 'the state' as messily and spatially renegotiated enables particular insights into activisms where the state/non-state binaries are contested (including distinctions between the state and civil society, see Jessop 1990, Andrucki and Elder 2007). Foucauldian conceptualisations of power have questioned the possibilities of 'standing outside' (in this case 'the state') and offering resistances from these positionings. Rather we are complicit, even through (queer) modes of opposition (Oswin 2005). As Cooper (1995) suggests, activists and the institutions they oppose might not always be as distant from each other as they are often portrayed (see also Cooper and Monro 2003). In considering the possibilities of LGBT politics within and with the state, we counter the presumption of coherent and knowable hetero/homonormativities that unite an entity called 'the state'. Instead 'understanding the state as a multifaceted, often contradictory entity makes it possible to think about, and engage with, the state in ways that extend beyond blocking and opposing' (Cooper 2011: 5).

Oppositional politics are key to queer anti-normativities that are often counter-posed against liberal bureaucracies, with assimilation placed in opposition to radical reconsiderations. For some, the emerging framework of civil rights that incorporates, rather than demonises, LGBT people has led to deradicalisation (Sears 2005). Critical (queer) discussions of legislative reforms can seek radical agendas that stand in opposition to new normativities. For example Stychin (2003) contends that those who contested reforms such as equalising the age of consent, began to be less interesting for scholarly critique, as the voices of reformers became 'increasingly hegemonic' (p. 26) at the turn of the century. The rise of queer activism is in part a response to what is seen as the mainstreaming of lesbian and gay politics (see for example Brown 2007). It creates and names 'new' enemies or opponents that can be opposed and resisted, contesting what are read as dominant social norms and assimilation. Naming an enemy or opponent (be that religious right groups, governments, blood donation service or the gay scene) is central to creating oppositional politics. Such tactics and conceptualisations produce an us/them binary that can be empowering and can challenge norms, discriminations and exclusions. However, they can also shut down other forms of activism and may not recognise the ways in which activisms can be simultaneously complicit and resistant (see Oswin 2005).

Writing from the USA perspective, Sears (2005) argues that, as civil rights meets the goals of many lesbians and gay men, there is a depoliticisation of certain forms of sexual politics. Yet he contends sexual emancipation has yet to be achieved:

> This demobilisation leaves many queers out in the cold. The consolidation of lesbian and gay civil rights has tended to benefit some more than others. Those who have gained the most are people living in committed couple relationships with good incomes and jobs, most often white and especially men. (Sears 2005: 93)

He argues that important gains in sexual rights have been made during a period when the (USA) political mainstream has shifted to the right, with significant cutbacks to the welfare state. Such USA centricity is exemplary of how 'capitalist states' and 'world cities' can be universalised to all, with the creation of particular models that fail to acknowledge and account for their locations (see Brown 2012, Robinson 2002, 2006). Other authors (some in the same context) have sought a more nuanced approach, 'recognising the hegemony of capitalist power relations while attending to the always contingent nature of these relations' (Andrucki and Elder 2007: 90) as well as questioning the portrayal of recent legislative changes as part of 'some grand (neo-liberal) conspiracy' (Cameron 2007: 1).

Celebrations of 'the World We Have Won' in relation to equalities and cultural changes (Weeks 2007) contrast with narratives of deradicalisation and leaving 'queers out in the cold'. Here scholars point to the supposed progress that has been made, particularly for lesbians and gay men in urban areas such as London. For LGBT groups, individuals and activists in the UK, there can be little doubt that the legislative change of the early twenty-first century was perceived to alter cultures and lives, such that engagement with LGBT issues improved in this time. Deradicalisation can be read through the assumption that sexual and gender liberation has been 'won'. Yet, there are dangers in failing to acknowledge that only some have 'won' (McDermott 2011).

We seek to follow those who contend that both oppositional/radical/resistant and assimilationist policies are necessary and can work in tandem (Rayside 1998, Weeks 2008). Such multiple strategies can be used in ways that recognise the possibilities of working with state process and other institutional forms of power, without negating the possibilities of other social worlds and other ways of being sought and created through radical agendas (see, for example, Brown 2007). Yet, as Rayside (1998: 5) suggests, working within the limits of the state can be seen as being at odds with the 'transformative impatience' of some activists, as regulating and normalising forms of activisms and desired goals. Thus, while most mid to late twentieth century radical gay agendas in the Global North incorporated legislative reform, the mainstreaming of lesbian and gay politics did not remove 'confrontational styles or radical visions' (Rayside 1998: 5). For Rayside, there is a spatial and temporal contingency to lesbian and gay politics that works across assimilation/radicalism:

> In some circumstances access and influence will depend on a group's representatives fitting in with the legislative or legal environment and playing mainstream politics by its rules, in others influence will depend on confrontation. (Rayside 1998: 10)

Thus, rather than being dichotomous or in opposition, activisms that work against *and* those that work with or within mainstream politics deserve attention, acknowledging the possibilities and problems of each in ways that are spatially and temporally contingent. Here, we couple the ambivalences of state actions and the rejection of hierarchies with the ways in which activisms evolve in relation to where they are created. The state here is not read as benign, something to be appealed to for 'rights' (see Spade 2011), nor however is it enacted in necessarily oppressive ways (Lloyd 2005). Rather, we seek to question these binaries and follow those who argue that enactments of the state can both act as firewalls to particular activisms and also 'function to produce creative and unexpected effects and articulations' (Andrucki and Elder 2007: 94, Bassi 2006; Cooper 2006). We now examine the manifestations of equalities landscapes where legislation, which once operated against LGBT activisms, shifted to work 'for' LGBT people, creating new forms of place-based LGBT politics.

Looney Left to Equalities Landscapes: Adversarial Enemies to Legislative Inclusions

In the 1980s and 1990s, the British New Urban Left came to power in local governments across the United Kingdom and sought to put pressure on the national Conservative government (Cooper 2004). Their policies and initiatives regarding lesbians and gay men were often in the front line of the 1980s culture wars, and these locally-based initiatives frequently encountered difficulties and retraction. Media attacks deployed the discourse of the 'looney left', and local progressive initiatives around lesbian and gay work received some of the most virulent critiques (Cooper 1994, 2006). There was policing of local authorities that sought to develop progressive lesbian and gay agendas, and those who did undertake this work often retreated in the face of media and local pressures. A determined Conservative government and financial issues led to what Cooper (2004: 154) terms 'general demoralisation on the left' (see also Cooper 1994, 2006). This retracted much of the innovative, creative and imaginative work that had been undertaken in progressive local councils during the 1980s, fuelled by a backlash to lesbian and gay equalities initiatives and organisational infrastructure, culminating in Section 28[1] (see Carabine and Monro 2004, Cooper and Monro 2003, Cooper 2006). The manifestations of this legislative and cultural 'venom', speaks to aggressive heteronormativities that requires and fueled activisms that depend on them/us, enemy/ally.

In Brighton (in the years before its repackaging as 'gay Brighton'), the 1990s were remembered as being played out through the public shaming of individuals for undertaking (paid) work for HIV organisations and Pride controversies that reiterated heteronormative orders. There were clear institutional 'enemies', with militancy evidenced by things such as the start of Pride in Brighton and Hove, which sought to address the national introduction of Section 28 (see also Chapter 8). In contrast, the election of the Labour government in 1997 offered a 'renewed boost of energy' (Cooper 2006: 927) and CMIT participants spoke of the hope that accompanied this electoral result. Gay Brighton in the early twenty-first century was understood as a 'different world' (Mabel) to that of the 90s, with the problematisation of prejudice and discrimination rather than homosexuality (Carabine and Monro 2004).

Whereas in the previous era 'the government of the time was extremely hostile to LGBT people' (Mabel), at the time of this research legislative change was seen as impacting on service provision. As Mary said, it provided 'a way of getting a rational reason for doing LGBT work' (see also Browne and Bakshi 2013). Brighton and Hove City Council (BHCC), one of the largest employers in the city, moved from fearfulness of undertaking LGBT work in the 1990s towards engaging in a number of initiatives to increase its LGBT friendliness from the turn of the century. In 2006–2009, it was ranked top out of all councils who entered the index of 'top 100 employers for gay [*sic*] people'[2] carried out by Stonewall (a national LGB charity). Thus, what used to be regarded as the 'loony left' by some came to be seen as part of the common sense of the 'politically educated':

> Nigella: Legislation is more a reflection of a certain type of social change because now in politically educated circles it's unacceptable not to embrace the equalities agenda. Whether you're left, right, green, whatever, equalities is where it's at. There has been a lot of change there and a lot of opportunity for LGBT political activism within that.

> I think this is a reflection that, within political circles it's not acceptable, but within communities it still is and within society it still is and homophobia is still rife within society. Yet we have all this legislation to protect us. The

---

1   Section 28 of the Local Government Act (1988) made the 'promotion' of homosexuality by local authorities illegal and labelled same sex families as 'pretend' and 'inappropriate'.

2   This 'workplace equality index' measures the 'gay' credentials of an employer, and ratings are calculated taking into account factors such as advertising in the 'pink media' and supporting a gay and lesbian employees group (Stonewall 2011).

politicians think that they can't be seen to be homophobic therefore they should have a whole lot of legislation in and does it make a difference? Well, no not really [laughter] homophobia's still rife and it's still really, really hard being LGBT whether you're monied with the pink pound, living in a penthouse flat in Kemptown, or not, we're all victims. We're still not seen as proper human beings by people who think they are proper human beings.

While becoming part of those that are acceptable can be seen as regulating and normalising, this must be tempered against the unease apparent in this narrative regarding the veneer of social change and continuing experiences of being a less than 'proper human being'. Nonetheless, the normality afforded by legislation was seen to lead to apathy (Sears 2005). There was nothing that could rally an 'us' against a 'them' and interviewees lamented 'resting on ones laurels' (Joseph) where there was work that still needed to be done:

> Persia: I have a sense that it's this kind of dangerous point of [view, that] everything is okay now. So we can relax and let go. I think in our society that such an inherent bias against LGBT people as a whole. We must always keep the eye on the ball. I've got a feeling sometimes in Brighton there's complacency, 'Hey, we're OK now'.

Peter contended that understanding Brighton as 'sorted' and 'successful' led to complacency. He did not perceive this to be the situation in 'other places' where 'there's an acknowledgement from the start that things aren't right'.

> Peter: My worry now is that we seem to have come to a plateau where I'm not convinced that we're moving ahead [in the way that] people perceive we are moving ahead. It worries me that so many people say the legislation's changed and everything's fine now and it's not. I think it's really difficult to do stuff here [in Brighton] because everybody thinks it's sorted. So you're fighting that prejudice straight away of council officers, who are middle class and very liberal, being mortified because anybody would challenge that it wasn't sorted. That for me has always been what I see as the problem. Whereas when you go to other places there's an acknowledgment from the start that things maybe aren't right and they have to try a little bit better for the gays. I think we've come full circle where the statutory authorities have become very complacent and they are saying, 'oh, we're very successful'. (Consultation event, April 2008)

The difference and protection of the 'Brighton Bubble' worked against political action that was based on oppositions between them and us. Those at the coalface of LGBT work, as well as those continuing to experience prejudice and abuse because of their gender and sexual identities, were very aware that more change was still needed. Yet locating 'the adversary' that would unite LGBT people was not straightforward, especially where legislative gains were being made:

Matthew: With most activists – the role of the adversary is key so you have something to fight against. So it's a sustained process. You have these opponents and elites that you can always rally against but when you start getting your civil partnership and your non-discrimination in accessing goods, facilities and services, and all these really important pieces of legislation, it's a bit like, 'Well, what am I angry about? I've kind of got everything that I've really wanted'. Of course homophobia exists, discrimination will still persist but whenever the big pieces of law are in place then it's what motivates people to get involved.

Faruk: When you're combating homophobia it's actually a really, really difficult thing to do. How do you fight homophobia? I have no idea how to fight homophobia. I can fight homophobic laws or I can fight hate crime, but homophobia is this much more difficult thing to [fight]. (Final feedback meeting, December, 2010)

This exchange at the final feedback meeting of the CMIT research points not only to the emotions of activism (see Wilkinson 2009) but also to the tangible nature of opposition, seeking something to fight. The 'inherent bias' that Persia discusses, and abstract concepts such as homophobia, need to be made tangible in order to be resisted. Activism can then be built on fighting homophobic laws, hate crime or other identifiable adversaries, which can form a purpose and generate political action. With the disappearance of a unifing agenda, under which we can not only unite but also coalesce as LGBT with shared experiences (see Browne, Bakshi and Lim 2011), there were worries about a fragmenting of LGBT politics:

Nigella: There is a real danger of fragmentation, that we'll end up with a grouping that's called something like 'marginalised LGBT people who are campaigning around LGBT issues' with this idea that there's an uncaring, L and G and B grouping out there who just won't get involved in fighting the battles. They're not taking on that the battles belong to all of us, including those people who choose not to fight. I hope it doesn't go that way. I'll be doing what I can to make sure it doesn't.

[There is a] lack of political thinking in Brighton that they don't seem to need to know why it's there [and] if you don't examine the politics of it, then you get left with all this baggage of unresolved tensions and ineffective working, cos you don't know what it is you're really fighting for.

Nigella worried about the effects of the fragmentation of solidarities and alliances between LGBT people in the reconstitution of LGBT politics in a city where those who were ordinary might be 'uncaring'. The already fluid grouping of LGBT people (see Appendix 2, Chapter 5) can be further divided by who fights for what in terms of political activism. Just as LGBT lives are heterogeneously created through place relations, so too are responses to sexual and gender inclusions.

Unresolved tensions and ineffective working were also part of the reconstitution of LGBT political thinking. While recognising LGBT politics as fragmented, Nigella argues for Brighton-based relations that should consider politics in terms of creating belonging for *all* LGBT people. This idealised vision builds in and from expectations of the city that fail to be achieved and yet continue to be hoped for and worked towards ('I'll be doing what I can to make sure it doesn't').

With legislative reform in the UK, as well as across Europe (see Stychin 2003), the enemy of a heterosexualised state was altered, and was arguably less visible and more fluid. Oppositional politics become less viable where tangible moments of 'homophobia' are less easily locatable, particularly in gay Brighton, but 'of course homophobia still exists' and indeed as this research shows biphobia and transphobia were key concerns (see Chapter 5, Browne and Lim 2008a, 2008c). Yet as Nigella alludes to, and as this chapter will now show, this did not mean that there was no evidence of activism in gay Brighton. On the contrary, activism could be located in multiple sites. This included partnership working with, and within, those insitutions once considered 'enemies',[3] namely the state as manifest through local government, the police and the public health services. We term this *ordinary*

---

3   This is of course not to say that other forms of activism were not present. For example, Queer Mutiny in Brighton is a branch of a broader UK-based movement and it has existed on and off since 2005. In 2010 it had weekly meetings/meet ups and organised on/off social, educational and political events. It understood itself within anarchistic and anti-capitalist terms. Drawing on transnational (arguably Americanised) understandings of queer as relating to gender and sexual fluidity, as well as the questioning of gender and sexual categories, the authors also set themselves up against 'LGBT space' in which they reported experiencing 'festishisation' and 'marginalisation' (Queer Mutiny 2010b). The Queer Mutiny manifesto (see Queer Mutiny 2010a) sought to 'eliminate heterophobia in the LGBT scene as well as the need for ghettoization in the gay scene', argued for self-organising, educating and critiqued 'LGBT culture, not the people who identify with it'. Although there were crossovers with those working 'with' the state and those who engaged with Queer Mutiny in terms of addressing state apparatuses, the LGBT activism in Brighton during this time often focused on working within rather than against. Working within was often focused on inclusion and addressing directly the needs of the most marginalised LGBT people who are also usually engaged with some form of state provision – health, housing, police, social services and so on – by engaging with those sectors and services. LGBT politics in the city discussed in this chapter could arguably be considered 'queer' in broader terms, but not necessarily anti-capitalist or anarchist (see Brown 2007 for examples of anti-capitalist queer activism).

The specificity of 'Brighton' and the local politics of the city was not apparent in the rhetoric or publications of Queer Mutiny, whose zines focus on (mainly US-based) international gender/sexual theory and discussions (see also Brown 2012). Perhaps what has yet to be explored is what Brighton 'queers' in ways that are not accommodated by globalised queer thinking and national movements. Considering politics as locally manifest and networked places as important could be an interesting queer activist and academic exploration.

activism where activists worked towards making LGBT people ordinary instead of sexual and gendered others.

Ordinary Activism: Partnerships and the Possibilities of Working With and Within

In the first decade of the twenty-first century, partnership working between LGBT communities and 'the state' in Brighton meant that it became difficult to segregate and target 'enemies'. This was not because there was an absence of evidence of abuses and discriminations and other violences from those who provided state services such as the police, local government and others (see Spade 2011). Rather, LGBT activists addressed these and other LGBT issues through partnership working with the understanding that Brighton was failing to achieve what it should and *should be* better.

'Partnership working' was not new or unique to New Labour's approach (see Rummery 2002; Craig and Taylor 2002),[4] but it did draw on an expanded understanding of 'social exclusion'. This moved from poverty and deprivation to a broader focus on marginalisation and integration, and extended the remit of the community and voluntary sector from delivery to strategy and policy development (Craig and Taylor: 2002). Gender and sexual difference was encompassed in new definitions of social exclusion. In Brighton the accepted LGBT population figures meant that services such as housing, police, health and others were legally required to work in partnership with LGBT communities. In establishing and developing LGBT partnerships, certain LGBT people and groups became part of the 'ordinary people' who should be involved in democratic processes, rather than excluded from them, as threats to the heteronormative state.

Partnership initiatives included a wide range of stakeholders such as private sector businesses, public service providers, service users, commissioners and

---

    4   The participatory democracies that underpinned partnership working can include partnerships between statutory and community/voluntary sectors (as well as with private and business sectors). Partnerships were seen as representing a 'third way', 'distinctive from both centralised bureaucratic hierarchies of Old Labour and the Market of the Conservatives' (Powell and Glendinning 2006: 1). Under New Labour, there was extensive development of partnerships that were supposed to increase democratic accountability and involvement. This sought to address the disengagement with formal politics and to tackle apathy by enabling people to be engaged and involved in making a difference in terms of social justice and inclusion, going beyond a concern with representative democracies to 'active citizenships', that is, citizenships that come with specific responsibilities (see McGhee 2003, Newman and Clarke 2009, Richardson and Monro 2012, Rummery 2002, Tett 2005). Community, civil society and third sector/NGOs were presumed to contain 'ordinary people' who could be called on to partner with/participate in new assemblages of power (Newman and Clarke 2009: 46) and in theory at least, to come to mutual understandings and shared goals (McGhee 2003, Powell and Glendinning 2006).

those who were seen as experts in a variety of ways. Here we focus on UK welfare that was delivered through intersections of public, private and community and/ or voluntary collaborations (Powell and Glendinning 2006). In Brighton LGBT individuals and groups were incorporated into policy planning, training of staff and consultations, alongside provision of financial support for certain LGBT groups in the city. Activisms were possible 'from within' and changing 'enemies' into partners, who worked officially and openly with gender and sexual difference to cater for LGBT people. This made it difficult (and at time unnecessary) to oppose public sector providers or politicians either through lobbying efforts within, or angry protests outside. This contrasts with Cooper's (2004) research which noted that in the 1980s local government staff felt blocked in their abilities to engage with lesbian and gay work and operated in informal and unofficial ways.

It is possible to read LGBT activisms in Brighton through partnership, cooperation and working together *for* LGBT people. Thus, where enemies were not easily located, or popularly recognised as such, LGBT politics were not necessarily absent, normative or indeed ineffectual, nor were the structures of institutions only or necessarily normalising and regulatory. Instead, LGBT politics were found inside as well as outside the gates of power, and tackled prejudice against LGBT people from within state based public organisations with the support of those organisations. This contrasts with other eras and places, where direct injustices perpetrated by influential state organisations provided a focus for activisms or lobbying efforts, and working within was not institutionally supported, either locally or nationally (Cooper 1994, Cooper and Monro 2003):

> Lee: If you look at Stonewall [1969 riots] that's one of the big defining moments in history and that kind of pretty much sets an idea of our relationship with them (the police)[as] something that's combative. I think it's healthy to be critical of things where you see they aren't working properly but I would say that actually we've moved on a massive amount since then.
>
> I'd always envied people who'd grown up in the 80s when Section 28 was going and the rest of it, purely because they had something to rally against, and there's a very easy way into being an activist there. Whereas now as things have become more professionalised there's not really that room for activists anymore I don't think. Actually we look for these activists constantly. Because I've been on the inside of statutory organisations for a long time I never considered myself to be one of those activists. I was always looking for these people who would come and do these amazing innovative things and looking to help them out if they materialised somewhere. And then the conclusion I came to was that actually the activists are us. Actually it's [name] and it's [name] and it's all these people who are already doing LGBT specific work.

In spite of the recognition that the 1980s and 1990s were difficult times for LGBT work within local government and elsewhere (see Cooper 1994, 1995, Monro and

Cooper 2003), this time was looked upon by some as a 'golden age' of political action where the enemy was clear and supposedly united sexual (and gender) dissidents in their fight against a known and identifiable heteronormative enemy. For some, activism did not exist in Brighton in the first decade of the twenty-first century, because this type of oppositional politics uniting 'us' was not visible in events such as Pride (Chapter 8). However, where there is a collective desire to work for and with LGBT people (who were read as a 'key part of the local community') other forms of politics may be created. Using ordinary activisms, as activisms that can seek to make LGBT people ordinary in territorially specific ways, LGBT work in terms of attending meetings with statutory services, doing policy work, working with public services and so on can also be seen as activism.

Ordinary activisms could be said to operate differently to earlier lobbying and protest efforts, moving from fighting, to partnership working with 'the state'. Activisms that entered through 'open doors' and were part of policy decisions and service provisions were viewed by some as 'fitting better' within the contexts of the first decade of the twenty-first century in Brighton. The term 'shouty activisms' was used for oppositional politics rooted in anger at injustices and located in histories of not being heard or listened to. Although these shouty activisms were important and empowering for some LGBT activists, they also stood in contrast to 'new' models that sought engagement through claiming rights and citizenships:

> Mabel: I can remember one time somebody slagging off a black person that I know because she's so loud and I felt like saying, 'Why is she so loud? Why does she shout?' It's so obvious: if you're not heard, you shout. I don't want to have to fight, but I've had to fight all of my life. The team [at an organisation] are trying to model a kind of 'we have our rights, as citizens, we have rights'. We don't have to fight and be aggressive. As I've got older I have learned to calm down and compromise more. I see this as maturity, not sell out!

> I think the police, NHS [National Health Service] and the PCT [Primary Care Trust] are full of great individuals willing to help us and doing their best. My perceptions are that organisations are like tankers out to sea, they take miles to change direction and we, like a fleet of little tugs with tenuous ropes attached, have to pull them gently and along those miles slowly in the right direction. I do not believe in any circumstances that it is right to forget that the crew of these monoliths are individuals, the vast majority of whom are principled and of good will. We must work with them not against them. It is no good shouting at them across the waters and ordering them about or giving them maps of where they should sail. We need to enable them to help us within the constraints imposed on them by multiple factors. This means throwing them our little lifelines and pulling with compassion.

> I think our model of involvement needs to go beyond activism based, as it often
> is, on protest and an inability to move beyond victimhood.

Mabel sought a model that 'moved beyond victimhood', yet recognised that 'if you're not heard, you shout'. In her example racial exclusions reproduced the figure of 'the angry black woman'. Maturity within the LGBT sector was characterised by not fighting and not being aggressive, coupled with the statutory sector 'changing direction'. However this was not easy or simplistic. For Mabel there was no point to 'shouting at them across the waters', rather we needed to 'pull with compassion', understanding the monoliths and restrictions of the organisational cultures in which individuals operated. Mabel's description of compromise with organisations who are slow to change but basically benign may have been necessary in order to operationalise her cooperative stance. We understand this as a form of LGBT activism with import in working for marginalised LGBT people from within, without negating the importance of shouting when you are not heard.

Nonetheless, as we will show below, her narrative at times contrasted with other people's perceptions and experiences of working in or with services, as well as critiques of citizenships and rights based claims (see Stychin 2003, 2006, Newman and Clarke 2009, Richardson 2004, 2005, Spade 2011). There are problems associated with a normative 'us' that works for particular LGBT lives, in ways that can normalise and regulate LGBT activisms within partnerships that require compromise. Monro (2005) points to the absence of bi and trans work in local government in the 1990s. In Brighton, bi and trans people continued to feel erased in the first decade of the twenty-first century (see also Chapter 5), illustrating the 'limited understanding brought to sexual orientation' and the ways in which local authorities resisted change (see Cooper 2006: 938; Richardson and Monro 2010).

Discussions of homonormative politics can place queer/radical/liberatory activisms in contradistinction with assimilation and equalities (see Duggan 2002, Puar 2007, Sears 2005). However, in contrast to discussions of the homonormativities that can elide and exclude differences amongst LGBT people, statutory services in Brighton in the latter part of the decade were seen (in limited and partial ways) to not only undertake bi and trans specific work, but also to cater for other marginalised LGBT groups (see Chapters 3 and 5 for some examples). Moreover, Brighton was seen by some as offering a queer politics that was not negated or homogenised through being 'institutionally LGBT', and indeed being 'institutionally LGBT' was not necessarily negative:

> Ellis: On the one hand I think it's really important that we're institutionally
> LGBT and that means it's really important that we have a scene and it's really
> important that the private sector and all the sectors, all the institutions, are LGBT
> positive. On the other hand, what that hides is this issue of widening the circle
> of empowerment until most people are kind of inside it or can aspire to be inside
> it one day, if they have a civil partnership or they want a family or they want to

own a business or whatever. But [what] if you're not interested in that, if you actually are different in other ways, or if you have multiple identities that don't make it easy for you to step into that circle? So on the one hand, I think it's really important to be institutionally LGBT, but on the other hand I think there also needs to be a real churn of lots of different kinds of competing and contrasting agendas that are much more diverse than that.

Brighton does sort of have that. It's resisted that de-queering. It's still diverse. It's still got a lot of people who, for one reason or another, aren't bought into the idea of social [inclusion] in that they want to end up married with 2 children or a civil partner with two children. Things aren't that easy for people. There's a lot of socially excluded people and there's a lot of kind of interesting discourse around how gender and sexual identities intersect in Brighton and Hove because it's got such a brilliant trans community.

Although working with or within 'the state' can be seen as 'selling out', such work is not necessarily (homo)normative. By this we mean that it does not have to regulate diversity and difference, while making LGBT people ordinary through widening the circle of empowerment. It was clear throughout this research that empowered LGBT people, who were listened to, used their position to augment work with marginalised people within the LGBT grouping, for example with trans groups, bi people, homeless people and other 'queers'. Place was important in orientating LGBT communities as 'the largest minority in the city' and this empowered some marginal LGBT people. Thus, partnership working and agitating to ensure legislative duties were fulfilled, does not necessarily 'de-queer' LGBT politics or leave 'many queers out in the cold' (Sears 2005: 93).

The argument for working from within as a form of LGBT activism does not negate the uneven ways in which 'inclusion' agendas and professionalisation LGBT groups operated (see Newman and Clarke 2009, Ward 2003). What has been termed 'second generation' LGBT activism from within has been seen as fraught and problematic, supporting institutional powers in part through the professionalisation of activism (Cooper 2006, 2009):

Ted: The LGBT sector became far more professionalised. We've got a lot of gains, but lost stuff as well. I think the stuff that we've gained is about the stuff in the non-LGBT sector, so about our issues being integrated far better in non-specific LGBT organisations. We have some amazing LGBT specific organisations doing a lot of very good stuff and run very well. It seems that a lot of the work that is happening is through organisations and that's very useful in some ways because it means those of us who are part of those organisations like to think that we have a good idea but I can't help feeling that there are a lot of issues out there that just don't get covered because the vast majority of us in those services have belonged generally to similar kind of groups.

Social subjects are made thinkable in particular ways, with some people not recognised in ways that can be catered for (Butler 2004, Carmel and Harlock 2008). Where communities need to become 'knowable' in order to engage with partnership working (Carmel and Harlock 2008) those who remained unrecognised were not able to 'get covered'.

The losses to LGBT activisms arising from working with the state were not confined to a lack of recognisability or to those 'excluded' from the processes of engagement. Cooper and Monro (2003) noted the constant restructuring of lesbian and gay work at the turn of century, which meant a change in roles and personnel attending to lesbian and gay equalities issues made local government 'fragmented and discontinuous' (240). Thus, although community groups can be read as unstable and indeterminate (Newman and Clarke 2009), in the CMIT research it was the fluidity of public sector workers that was 'the problem'. For some the commitment, investment and training of key workers in public services was often lost, as they 'move[d] on', replaced by those who once again needed to be 'educated' (Jason, consultation meeting April 2010). This pointed to the longevity of key LGBT actors in contrast to ever changing public sector roles assigned to individuals. Moreover, over the course of this research, organisations sent people to meetings 'who didn't know anything' with regards to LGBT issues and were not properly or appropriately equipped to engage in meaningful partnership working. The resulting conflicts did not always focus on their lack of preparedness to undertake this work:

> Charlie: There was this one meeting with the PCT [Primary Care Trust] and the council with members of this trans activist group, and the meeting broke down. The PCT was, as usual, obdurate and unreasonable and sent in someone who didn't know anything about trans issues. Every single meeting they send in some other new idiot who doesn't know anything about trans issues and just goes 'I'm sure it's all sorted', when they don't understand.

> There was all this frustration flying about and members of different kinds of trans communities started having a go at each other and it just all kind of broke down. I had to go back to work after this meeting on the bus, [and I overheard statutory representatives from the meeting saying] 'oh well we don't really need to worry about this, we'll come back to the table when those kind of internal divisions are dealt with'.

> It's just so frustrating when it happens. It's so frustrating to watch people go 'oh okay well we don't need to do that bit of work then'. Because it damages equality for all of us, when to most people who don't understand the specificities of our different communities, we're all LGBT. It damages all of us when we can't get it together to put our differences aside. When I say differences, I don't mean difference as in diversity, I mean the history. The painful histories people had interacting with institutions, getting what we can from them as individuals or as

groups and then having to deal with risking all of that to get something better for all of us. It's a difficult process and I think it's really scary and hurtful. We all wanted the same stuff when we went into that room and we came out feeling that we didn't and that difficulty has stayed with me. It's really disheartening to me.

The imperative to consult in Charlie's narrative was fragile and easily disturbed when internal divisions were aired in forums created with public sector workers. Even though differences were recognised between L, G, B and T, there were often calls for representatives to present a homogeneous view of 'the community'. As Craig and Taylor (2002: 140) argue, while communities are expected to 'squeeze their views through one or two representatives, public sector agencies (and business partners) rarely experience the challenge of representation that communities struggle with'. For Charlie differences should have been 'put to one side', because most people 'don't understand' and to them 'we are all LGBT', such that these arguments 'damage us all'. The demand for a specific form of representation disrupted activists working with public sector partners. Rather than exposing the complexity of LGBT groupings and the diversity of possible approaches to cater for LGBT people, these scenarios were used to stagnate work on LGBT issues. The 'them' and 'us' opposition cannot be blamed for this breakdown, nor can the homophobia, biphobia and/or transphobia of the public sector, rather the 'us' was not playing by 'their' rules of engagement and 'listening'.

Public sector bodies can both know and control the 'rules of the game' and thus create complex institutions, regimes, cultures and barriers which prevent people 'from fully exercising their power in local partnerships' (Fuller and Geddes 2008: 274). For Craig and Taylor (2002: 134) this is due to public sector cultures that 'are so ingrained that power holders are often unaware of the ways in which they perpetrate unequal power relations through their language and procedures'. Rather than known procedures, it was the *lack of knowing* that led statutory sector individuals to presume everything to be 'sorted'. The painful histories of engagements were played out because of 'obdurate and unreasonable' engagements at a strategic level that practices forgetting (Cooper 2006, Monro and Cooper 2003). This can cause tensions between them and us but, in highlighting differences within the 'us', it can also result in internal fighting. Such strategies illustrate how state agents can resist LGBT work, while providing the illusion of working with the LGBT community (see Cooper 2006). Thus, discussions of normalisation, regulation and the problems of 'working within' are important when examining ordinary activisms created through institutional spaces. We are arguing, however, that ordinary activisms cannot be *reduced* to normalisation and assimilation. Instead, we move on to illustrate the multiple ways in which ordinary activisms operated across radicalism/assimilation. We do this through an in-depth discussion of Spectrum, a local LGBT community group that worked between resistance and collusion, deploying oppositional tactics alongside new forms of partnership working.

Thorn in the Side?: Spectrum and Ordinary Activisms in Gay Brighton

Carmel and Harlock contend that during the period under discussion, diverse charity, community and local groups were recast, creating 'the community and voluntary sector' and this 'institutionalised the idea of a single, unified social actor, with whom government and public sector could deal' (2008: 158). During the New Labour era, there was a normalisation of professionalisation and conformity to business models amongst many categorised within the 'voluntary and community sector' (Carmel and Harlock 2008, Cooper 2007, Monro 2007). Thus, for some, partnerships are seen as an extension of neo-liberal governance with community groups and community representatives becoming 'agents of government as much as advocates for their localities' (Fuller and Geddes 2008: 275, Carmel and Harlock 2008). These are seen to depoliticise NGOs and community groups through the imposition of state practices, rules and delivery targets onto previously independent organisations. Moreover, for some, the localisation of partnership working is an issue and fragmentary politics are weak, 'contained within limited horizons beyond the neighbourhood or locality, which makes it unlikely to develop a broader contestation of neo-liberalism' (Fuller and Geddes 2008: 277). While professionalising NGOs can be interpreted as 'reproducing dominant forms of knowledge production, neo-liberalising social movement activism and reducing the sphere for advocacy and democratic politics', partnership working also offered the potentials of being 'exploited by marginalised groups seeking entry into policy discourse' (Grundy and Smith 2007: 295). Monro (2007: 12) contends that for lesbian and gay [*sic*] groups, the adoption of professionalisation offered legitimacy for community activism. Here we use one community and voluntary sector organisation to show that LGBT community and voluntary sector groups could deploy multiple tactics such that working with the state did not necessarily mean becoming 'agents of the state'.

In gay Brighton a key element of the partnership working that drove LGBT politics in the twenty-first century was Spectrum. Spectrum defined itself as:[5]

> Spectrum is Brighton and Hove's Lesbian, Gay, Bisexual and Transgender Community Forum established in 2002 to provide infrastructure and community development support to LGBT communities and promote partnership work and community engagement in the planning of services and policy. (Spectrum Statement of Purpose)

From 2002 to 2011 (when it closed due in part to funding issues and the change in political climate), Spectrum developed an identity, position, relationships and working practices that enabled it to facilitate the development of structures and relationships that promoted cooperative engagement with LGBT communities.

---

5   This was part of Spectrum's statement of purpose published presented in 2008, which is no longer available online or in hard copy. Copies can be obtained from the authors.

Working with, rather than campaigning against, Spectrum sought to influence and work with agendas of statutory services. It organised events and large scale LGBT public forums, where service providers and others reported to LGBT communities; set up working groups around particular issues (such as safety, drugs and alcohol and housing); and worked with service providers and others to develop empowering consultative modes of engaging with LGBT people. These were imperfect, as we will see, but they were also important in working for some marginalised LGBT people whose voices would otherwise not have been heard.

Spectrum was proactive in seeking to support representation of all LGBT people, including marginalised and disenfranchised groups, 'working with individuals/groups to help them establish, increase capacity and effectiveness' (Spectrum Statement of Purpose). Spectrum's focus was often on including the marginal and disenfranchised, those who did not fit into gay Brighton, seeking to work for their inclusion. As a result it was also often on the fringes and connected to hidden LGBT people who did not 'fit' in the gay capital. It occupied an unusual position in not being a campaigning group, yet also not acting as a service provider. Instead Spectrum facilitated engagements with LGBT people.

Spectrum supported public sector bodies to meet their statutory duties to consult with their LGBT local communities (see Chapter 7 for a discussion of these duties). It undertook this role in ways that worked against some of the pitfalls of community representation, such as the reliance on one person and the absence of a community to be consulted with (Craig and Taylor 2002, Fuller and Geddes 2008). Despite being funded by public sector organisations and seemingly situated within the professionalisation of LGBT groups, Spectrum's workers challenged public sector cultures that sought 'tick box' approaches rather than meaningful consultations with LGBT populations. In this way Spectrum was empowered to rework state resources:

> Dana: They [statutory sector] wanted to be seen to be doing what they were meant to be doing. So they wanted to tick the LGBT tick box and I wasn't prepared to collude in that. I would do the schmoozing to a certain point but then I'd do the sort of public shaming if it was necessary.
>
> Interviewer: Can you tell me about public shaming then, how did you do that?
>
> Dana: Well if they came along to meetings saying what the council had achieved and I knew they were lying I'd say, 'That's not true, you're not doing that and this is what you need to do.' I would get angry.
>
> [I felt] anger from the statutory people who felt under attack and felt as though they were cornered because they had to do the LGBT [lesbian and gay] thing. I did feel we had quite a lot of power at that point because they'd all signed up to 'Count Me In' [see Chapter 2] so you could hold them to account.

Cooper (2004) noted those who work from the inside as activists are not necessarily respectful, adherent or predictable. Dana, the Spectrum worker until 2004, discusses the ways in which she rejected collusion and deployed activist tactics that sought to shame. The ability to be shamed when not undertaking LGBT work illustrated the importance of Brighton in constructing a new form of ordinariness where statutory organisations should, and indeed were required to, undertake LGBT work.[6] Thus, relations between partners were not always diplomatic or friendly, nor were the tactics used by LGBT groups such as Spectrum as they engaged with public sector institutions. LGBT groups did not have play by their rules and activists at times felt empowered to challenge these rather than working in collusion. It is clear that the imperative to consult and the desire to make change happen caused conflict and acrimony, as well as facilitating collaborative partnerships. Spectrum worked with public services and also acted as a 'thorn in the side' of those institutions. As such, Spectrum was both a partner and a radical organisation that disrupted norms, cultures and institutional practices.

In occupying a position where Spectrum dealt with the council, police and health services, LGBT voluntary and community groups and LGBT people, Spectrum employees worked across sectors. At times it brought diverse sectors together to address particular issues, for example housing, the dangers of civil partnerships for marginalized LGBT people. Spectrum was understood by some as having 'power' with statutory organisations because, at times, Spectrum workers and trustees were listened to and acted as quality assurance on work undertaken by the public sector:

Persia: Spectrum does have some power because it's now been given a position within the town. The council and stuff always need to have Spectrum's okay on things. This means it gives us power. I think Spectrum does have power, but it's the power of influencing and denying.

Interviewer: The power of denying?

Persia: That people can't do stuff without their approval, right? So a policy has to be good according to Spectrum, according to what they think. Spectrum is needed now to rubber stamp things and that is a real power. It keeps things in line basically.

Spectrum, for Persia, acted as a watchdog of the public sector that championed marginalised LGBT causes working with public bodies to improve their provision to LGBT people. Spectrum acted to keep 'things in line', seeking quality services for LGBT people, rather than settling for those that 'are not perfect yet' (Mabel). In this way, Spectrum's approach was both adversarial and working in partnership

---

6   This of course ties in with other work that has noted the productivities of shame (see Munt 2007, Probyn 2005, Sedgwick 2003).

with the statutory sector. Examining a group such as this then blurs the distinctions between assimilation/radical politics.

However, blurred boundaries and the lack of 'service delivery' were not always appreciated. For some it was problematic that Spectrum did not to have a recognisable remit or fit within specific structures:

> Natalie: Spectrum's become quite powerful, quite quickly. I don't see ourselves and [name of organisation] and [name of a different organisation] as being massively powerful, probably because our core remit is to deliver services in a way that's not the same as being in a campaigning role or being in, I don't know what the word is, but Spectrum's more of a community forum. So do we want power anyway? We kind of want to provide services for service users. So our role's a bit different.

> I do have some questions and doubts about what their [Spectrum's] role is, as well. But it would make sense that they would be in a more powerful position, because they don't deliver services. I don't really know, are they a community forum? Are they a campaigning group? Are they a community watchdog? And if I observe what they do, they seem to do all of those things. But for example, on their website it says quite clearly that they're not a community watchdog. But I observe [Spectrum] filling the role of community watchdog in a range of ways. So that leaves me feeling confused about what exactly their role is. And it kind of makes me wonder if they're confused themselves, because to say you don't do something and then do it doesn't sort of sit right with me. So I just feel a bit confused and then that leads to confusion for me in my partnership with Spectrum about what can I can reasonably expect of them, because how can I know that if I don't fully understand and feel clear about their role?

Natalie expressed unease with a lack of clarity and seemingly contradictory discourses and actions. These were not celebrated as fluid and malleable, enabling diverse and strategic practices; instead the demand for coherent, recognisable roles permeated through what LGBT groups expected of each other, in an era of professionalised partnership working (see Craig and Taylor 2002; Newman and Clarke 2009). Lacking such specificity, Spectrum's agenda could be responsive to external strategic events and not focus on the practicalities of the delivery of a specific portfolio of services or specific campaigns.

Spectrum's function served to hold public services to account, through a variety of mechanisms. Although these tools did not include protests, direct action or polarised press debates, fighting and confrontations were part of the tools deployed. However, in deploying resistant strategies as well as Mabel's tugboat tactics (see above), Spectrum was critiqued by other LGBT organisations and LGBT statutory sector workers and activists who felt that those organisations, which Spectrum was confronting, were 'doing their best'. They asserted that the block to effective working lay with those 'bashing' public sector services (see also

Browne and Bakshi 2013). Indeed confrontations were regular and the Spectrum worker in the latter part of the decade was perceived as difficult. This was in part because they refused to be 'the representative of the LGBT community', demanding instead more open and inclusive engagements. They also would not bow to the institutional pressures to conform or 'make do', demanding instead proper processes and procedures that were transparent, but took time.

Spectrum was not subsumed, co-opted or assimilated into statutory ways of working. It was both inside and outside simultaneously, operating between collusion and transgression. However, in not providing mass protest actions and in being tied financially to the public sector, Spectrum was seen to forego the possibility of occupying the role of an oppositional, activist organisation. There was a desire for a popular, well-known activist organisation – in the sense of an oppositional group that worked against particular (state) organisations. Participants argued that this was needed to complement work with and by statutory sector groups. The gay media in Brighton adopted oppositional positions, with individuals making decisions on who was right and wrong, them or us, as well as who could represent 'the gay community'. Joseph discussed the 'independent gay press' as the key form of oppositional activism in Brighton. This mechanism he believed held them to account. His role as part of this was, in his view, due to a void created by the absence of a specific activist group in the city:

Joseph: The independent gay press has had to do the kicking and then do the reporting and that's hard to do. There's a conflict there as well in doing that 'cause you're walking a very fine line as to being the bad boy one minute and then having to sit with the same people and be the nice boy. I can do that, no problem at all, but it's how people receive it, whether people are comfortable with you doing that.

Interviewer: How did you kick?

Joseph: How did I kick?

Interviewer: Yeah.

Joseph: The ... I ... well, I ... it was ... we just used the [media outlet] as the hammer.[7]

The use of the media outlet 'as a hammer' was seen as giving Joseph a significant amount of power in the city that lacks an icon for LGBT rights and instead in part operated through the fear of what might be 'pushed forward in our LGBT press'. However, as we have seen in earlier chapters and will discuss in Chapter 8, this

---

7   We did not 'tidy up' this quote, as we do for all other quotes (see Appendix 1), because how it was said is important.

presumed homogeneity also recreated hierarchies between the 'us' and the role of being a 'hammer' was used in ways that were perceived to be negative as well as positive. The development of coherent oppositional agendas can also recreate inequitable differences, marginalisations and exclusions.

There can be little doubt that the successes of partnership working in gay Brighton and inclusion of marginalised LGBT people onto key public sector agendas was in no small part due to Spectrum, its workers and volunteers. Setting up and maintaining relationships, while continuing to act as a thorn in the side, was perceived as setting gay Brighton's community sector apart. Nonetheless, as in other areas, 'suitable representation of all community interests is contingent on a few individuals' (Fuller and Geddes 2008: 273). Working within, as well as against, can be costly for those who represent community and voluntary sectors. Community groups can be seen to inequitably bear the underestimated human and resource costs of partnership working, 'with the public sector reaping most of the benefits' (Rummery 2002: 338, Craig and Taylor 2002, Fuller and Geddes 2008, Newman and Clarke 2009, Tett 2005). Fuller and Geddes (2008) highlight the 'burnout' of those involved in partnership working from 'long and unsocial hours spent as largely unpaid volunteers within the state apparatus' (262). Here we finish with two examples of the personal consequences of partnership working on LGBT people.

Working with state bodies, without being affiliated to them, could 'damage reputations', particularly when these state bodies continue to be the enemy. On the one hand, working too closely with 'them' may not be appreciated by those who see this as collusion ('coppers' nark') (Fuller and Geddes 2007, Sears 2005). On the other hand, it can be a form of power where some groups are being listened to and others are not being heard, or are being silenced:

> Joseph: I didn't realise it till I saw myself on television and I couldn't believe what I said in that it could have been a spokesperson for the council or the police. I'd just taken on the whole language that they use and not realised that I was doing it but I've been working with them for 40 hours a week and it had just affected everything I'd done. I just sort of got this whole new vocabulary. I then started saying to people 'if I ever say anything you don't understand just tell me' and the first person I said it to, said 'what's a strategy?' Because that's what I was talking and people didn't understand and that was an eye opener for me – that we were using language that people on the street didn't understand.

> When I was doing the AVU [Anti-Victimisation Unit, see Chapter 7] it damaged my reputation because people thought I was working too closely with the police and that caused me [an] awful lot of problems. People thought I was a coppers' nark and I knew too much.

As Monro (2007: 13) notes, hybridisation meant that 'discourses of mainstream organisations were adopted but altered by community organisations and vice versa'. Colluding with the (perceived) enemy in order to engage and participate in state safety institutions, Joseph found himself not only immersed in particular modes of discourse, but also transformed by them to the extent he was not understood 'on the street'. LGBT people are thus reformed, reconfigured and remade through engagements with the state in ways that they may not realise until abruptly confronted with the gap between themselves and other (ordinary?) LGBT people.

The direct 'attacks' that some LGBT activists experienced came not only from statutory services, but also disenfranchised LGBT people:

> Dana: I'd been there [in Spectrum] for I think three years. I was absolutely exhausted. I think there's something in the community about very high expectations of people inevitably failing. You inevitably can't live up to them. I think there's such disappointment because in terms of social exclusion locally it is so painful, because you know people are suffering so much. There's so much anger if you're not seen to be giving enough of yourself or doing enough, and you can never do enough and you can never give enough of yourself. It's pretty harsh.

> I think there is something about the impact on people's mental health of doing that sort of job because [man's name] worked 24 hours a day; he gave his life to it. I felt guilty when I was at Spectrum because I wasn't prepared to do that. I felt judged for that locally because [man's name] I suppose was the comparator.

Dana describes the demands that were placed on those who worked with the statutory sector, but 'for the community'. LGBT people who worked for community groups (and those in the public sector) were often expected to work long hours, to attend meetings and organise events during unpaid time and to undertake other jobs and roles as volunteers on top of their paid work. Dana and other interviewees noted the effect on the mental health of those who worked in this field and the costs of working within LGBT communities, as well as traversing the community and statutory sector as both an insider working in collusion and someone who is also outside. Internal conflicts were a key factor in the mental health impacts and burnout that Dana and others described. Dana left Spectrum, and other women spoke of similar trajectories associated with working with and for LGBT communities across diverse sectors. Explanations in the data of what went wrong often suggested individual abuses of power. However, Dana argued for socio-psychological rationales that were created by Brighton, as it played a part in the reconstitution of inequitable hierarchies between LGBT people:

> Dana: There are very, very strong personalities in Brighton and you've got to be pretty tough too. It was hard. People get burnt out in a really appalling way. It feels like people just reach the point where they're burnt out and hurt and rejected and disappear.

I think a lot of it was internalised homophobia. If you're a certain sort of gay man you do have that male socialisation which gives you a certain sort of power that it can be more difficult for women to claim. And the combination of that power and that anger about their own mistreatment over the years, which wasn't actually being articulated, felt quite dangerous because it felt very unconscious. It felt as though people were acting out all over the place and not knowing what they were doing so calling it something else. The viciousness towards lesbians locally, there was a real strain of misogyny that would sometimes be explicit that I'd hear about. They'd [particular gay men] be quite threatening – I mean [name of media outlet] publicly shames people all the time. I'd pick it up and I would just be dreading what it was going to say every time. I was waiting to be lynched in it you know. It wasn't pleasant, and then it would be concealed by a sort of faux warmth and flirtiness but you couldn't trust it. You'd meet [male names] and they'd be warm and they'd be friendly and they'd kiss you and they'd ask you how you were. They'd have a drink with you but you knew you weren't respected.

If you're LGBT, the more aware you are of what you've done with how you've been oppressed, the more political choice you have. The less likely you are to get burnt out or become a bully or be oppositional or whatever it is. You've got much more room to manoeuvre. I think there's sort of blind acting out happening a lot. I mean I did it as well; I wasn't exempt from that. But it means we just keep hurting each other. That's the horrible thing – we just keep hurting each other in the community.

For Dana, working productively within LGBT communities in gay Brighton, as well as with the statutory sector, requires more nuanced understanding of oppression and ourselves, to consider ways beyond forms of oppositional modes that seek enemies 'within' and can be self-destructive. Activists often were not exempt from 'acting out', such that there was, at times, little care for the self and others in people's experiences of LGBT activisms in the gay capital. The 'nastiness' experienced by LGBT activists did not only come from supposed enemies, but from within LGBT communities where new enemies were found and fought.

Newman and Clarke contend that the new assemblages of publicness, created through partnership working and other forms of participatory democratic initiatives, have a plethora of 'sources and resources' that are brought (in their terms, 'dragged') into 'complex, uncomfortable and contradictory alignments that produce unstable formations and may have unpredictable consequences' (2009: 175). Here challenging both discourses of the 'progressive politics of the public' and those that see assemblages as products of neo-liberalism that mask 'the decomposition of power and assuring the conditions of capitalism' (Newman and Clarke 2009: 175), we have shown that organisations such as Spectrum can offer radical as well as assimilationist potentials. Community and voluntary organisations are thus multifaceted and politically ambiguous and working in this way can be personally costly.

Conclusion: Hopes for Politics and Partnership Working

Taking into account the specificities of *where* activisms are manifest, place is important in exploring what counts as 'politics' and what activisms can look like. While it has been acknowledged that partnership forms are situated in their own genealogies (Newman and Clarke 2009: 68), we would argue that place is also an important constituent not only of partnership working but also of the activisms that are created through this. This chapter has shown that partnership working occurs in, is shaped by, and responds to the context of imaginaries of place and space as well as understandings of the goals and processes of such work. Gay Brighton played a part in the formation of political activism in the city. The successes of LGBT identity politics in Brighton created possibilities of working for change, collaborating with or employed within organisations that had previously been framed as 'the enemy' (although these institutions and individuals had undertaken gay and lesbian work in the past, see Cooper 1994). Taking place seriously is key to understanding how LGBT activisms emerged across the whole city of Brighton, including the formation of institutional settings.

Multiple forms of activism can sit at complex angles (Cooper 2009, see Chapter 1) to power relations enacted by state actors, powerful gay men and others who deploy resources to (re)produce inequitable relations. Where there is no need to 'forcibly enter ... into mainstream politics' (Rayside 1988: 14), there is an imperative to critically examine the limits of inclusion and the reproduction of equalities rather than liberations (Duggan 2002). Here we examined a place where legislative change and local policies demanded engagement not only with LGBT people represented through lesbians and gay men (Richardson and Monro 2010 2012), but also with marginalised and disenfranchised people who come under the category of LGBT. We did this through discussions of partnership working which questioned radical/assimilationist divides. In our research, oppositional and partnership initiatives were not neatly delineated, and individuals did not engage with these in a mutually exclusive manner, implying that moving beyond oppositional politics can mean engaging in both.

Our discussion of resistant and oppositional activisms continues through the rest of the book. Developing discussions of partnership working, the next chapter explores safety partnerships in Brighton, where genealogies of 'failed' relationships led to multiple activisms, including resisting the state and radically reconceptualising safety beyond reporting and criminal justice. This illustrates that although partnership working offered possibilities there was also evidence of tensions, normalisations and recuperations.

Chapter 7
Resistant Ordinary Activisms: Safe in the 'Gay City'?

Introduction

Critiques of equalities policies and legislation need to be supplemented with explorations of the messy ways in which state legislations are put into practice (Ahmed 2012, Carmel and Harlock 2008). [1] Indeed there is no reason to assume that partnerships with LGBT people and groups such as those introduced in Chapter 6 are consistently enacted (Powell and Dowling 2006), even where uniformity might be desirable in order to ensure that LGBT people are 'on the agenda' (Monro 2007). Political projects and governing strategies are not constant, rather they are 'assembled, condensing multiple political desires, diverse discourses and repertoires of governmental devices' (Newman and Clarke 2009: 179). The plethora of genealogies and histories of working with the state can diversely reconstitute the terrain of governance and how this is manifest through its disparate (spatially specific) enactments (Andrucki and Elder 2007, Grundy and Smith 2007, Newman and Clarke 2009, Tett 2005). Thus, the stated intentions of national and local government can be mediated and altered by those within partnerships, producing multiple effects and outcomes. Yet, Newman and Clarke (2009) warn that while spaces of indeterminacy are spaces of possibility, they are not necessarily progressive. Governance techniques 'have been reworked in unexpected ways' and specifically 'redeployed in the formulation of demands

---

1   Although we are dubious of strict structures that define and determine the 'factors' that make partnerships or participations 'successful' (see Chapter 2), authors have suggested more general aspects that could be variously applied, that might aid in the development of effective partnerships. Craig and Taylor (2002) argue that the key things needed for 'real partnership' include better information and communication; early engagement; effective resourcing; recognition of power inequalities; attention to the needs of minority communities; clarity about goals; compromise where appropriate and 'a recognition of the need for organisations within the voluntary and community sector to protect their own role and purpose' (142); and an understanding of the 'different contributions each partner can make and how they can best support or complement each other' (143). Tett (2005) contends that power relations between partners need to be clearly acknowledged, 'all partners have to be clear about their position in relation to marginalised groups' (13) and 'new ways of thinking about representation are required, with priority given to including those with the least power who are nevertheless the most expert in identifying the needs of their own communities' (13).

made by LGBT citizens on authority' (Grundy and Smith 2007: 312) and such work exemplifies the 'complexity of contemporary political' life. Grundy and Smith read this as creating 'the possibility for modes of resistance and contestation' (313). Building on these discussions, in this chapter we examine the complexities, ambiguities, possibilities and tensions of working in partnerships, with and within the state. We do this through interrogating an example of partnership working that was perceived to 'fail'.

In the national context of the New Labour government, safety was seen as key to creating a just society and as an important contributor to reducing social exclusion (McGhee 2003, 2004). Cooper and Monro (2003: 248) contend that at the end of the last century, 'lesbian and gay work became incorporated into the new anti-hate crime agenda', attributing this to the presentation of lesbians and gay men as 'victims'. The departure from criminalisation and deviance of some sexual and gender dissidents meant that LGBT people were instead seen as excluded, abused and needing protection (see McGhee 2003, Moran 2001, 2007). McGhee (2003) notes that hate crime is now associated with social cohesion such that hate crime is related to incivility, where 'the homophobes' and transphobes'[2] conduct ... is increasingly being seen as out of place in late modern society' (McGhee 2003: 355). New Labour's holistic ethos with regards to safety linked hate crime to the promotion of social inclusion and well-being, in particular through working for appropriate responses to hate crime (McGhee 2003). Thus, on the UK Government Home Office's own terms, homophobic hate crime and discrimination based on sexual orientation ran 'counter to the Government's [New Labour] avowed principles of equality and hindered the establishment of a safe, just and tolerant society' (Home Office 2000: 98,[3] cited in McGhee 2004: 370).[4]

----

2   Note the absence and erasure of biphobia.

3   This quote initially refers to the reform of the sexual offenses act to remove buggery and the consultation surrounding these changes; it then goes on to state 'differentiation in law can be seen as justifying discrimination and homophobia; neither of which acceptable in a civilised country' (Home Office 2000: 98, cited in McGhee 2004: 370). McGhee notes that this can be seen as part of the 'Europeanization' of the UK and resulting legislative changes.

4   While official bodies such as the police have moved from the persecution of 'homosexuality' towards working with LGBT groups (see McGhee 2003), there is a continuing need to 'address deeper social divisions that cause crime' (Yarwood 2007: 456). As Perry (2001: 3) notes, hate crime is a mechanism of violence that sustains 'precarious hierarchies'. Internal conflict between LGBT people also has implications for safety, and these are not addressed in a model of violence by straight people against gay people or a focus on stranger danger in public (read straight) space (see Warrington 2001, Valentine 1989, Yarwood 2007). Moran et al. (2004) contend that failures to move beyond straight/ gay, straight/lesbian conceptualisations of safety/danger are limited in how they engage with risk, safety and danger. In Brighton, biphobia and transphobia in scene/LGBT spaces were often overlooked and ignored, despite these being classified as hate crime. Safety was not experienced equally across or between lesbian, gay, bisexual and/or trans people, either

Key to establishing safety was partnership working between police, other safety services and the community and voluntary sector.

We begin the chapter by outlining some of the critiques of hate crime and criminal justice, using these to critically explore how imaginings of Brighton were active constituents in policing and safety initiatives. Safety initiatives were a context for partnership working between diverse sectors, and safety itself was contested with no shared definition or understanding of what constituted being 'safe in the city'. We next move to explore LGBT partnership working with statutory services, and highlight the histories and ongoing tensions of working with and from within. Forcing 'community consultation' (see Tett 2005) may have been a national initiative, but the intricacies of Brighton illustrate the importance of place in conceptualising the complexities of partnership working and ordinary activisms that included 'shouty politics'. The second section outlines the long and fraught histories of conflict, mistrust and failures that constituted LGBT safety initiatives in the city at the turn of the century. We show how spaces for partnership working remained fragile, uncomfortable and contested. Thus, despite national legislation and the dissonances between the ideals of Brighton and the realities for LGBT people (including the rates of abuse LGBT people experience on the basis of their sexual or gender identities) that were seen to spark activisms in earlier chapters, LGBT safety initiatives in Brighton were fraught and stalling in the period under investigation (2000–2010). Relations between LGBT communities, the police and the local authority-run Partnership Community Safety Team demonstrate the disparate way in which safety was enacted and partnerships operationalised and contested. The chapter concludes with a vision from Spectrum, the LGBT community forum, that outlines how 'safety' might be more broadly addressed in community and practitioner contexts suggesting tangible, if idealistic, ways forward. In this way, this chapter interrogates both partnership working and safety as key features of equalities agendas at the turn of the century, finding complexities both in how the state was enacted, and also how ordinary activisms were manifest.

Safe in the City?

> Audience member: Safety is absence of injury and that injury could be disease, mental illness, stress or physical harm, and safety should be absence of fear. We should be able to walk the streets of Brighton without fear. (Consultation event, April, 2009)

Safety is more than an absence of abuse. It is far more nebulous than this and relates to broader societal 'acceptances', feelings of safety, possibilities of enacting

---

in terms of the experiences of abuse (from mainstream or other LGBT people) or the effects that these experiences had on them (see Browne, Bakshi and Lim 2011, Herek et al. 1997, Clements-Nolle et al. 2006, Szymanski 2005).

LGBT identities in taken for granted, indeed ordinary, ways (Browne, Bakshi and Lim 2011). An absence of safety due to perpetration of abuse is damaging not only because of the personal cost to those who experience it, but also because the knowledge and awareness of specific incidents and experiences based on an aspect of one's identity is in itself a marginalising force. The criminalisation of hate crime and the recourse to criminal justice and reliance on state or police interventions have been critically explored (Cooper and Monro 2003, Hutta 2009, Mason 2007, Morgan 2002, Moran et al. 2004, Richardson and May 1999). Moran et al. (2004: 27) document some of the 'perils of pursuing emancipatory aims through law and order agendas'. McGhee (2003: 367) argues 'what are the alternatives ... paying the dues and taking up the responsibilities of the active citizen or remaining "hidden targets"?'. Moran et al. (2004) do not argue against criminal justice in lesbian and gay politics, rather, they seek to open the debate, perhaps beyond the dichotomy McGhee puts forward, to understand legislation and its enactment as offering both promise and problems. Yet they caution against the use of the violence of the law, noting that the 'emancipatory sexual politics against heterosexist violence is all too closely associated with social and cultural trappings of our unfreedom' (Moran et al. 2004: 172).

In spite of the legislative and cultural gains for LGBT people during the first decade of the 21st century, experiences of violence, abuse and discrimination emerged as a common experience for L, G, B and T people in CMIT, and one that we have argued lends coherence to the category LGBT (Browne, Bakshi and Lim 2011). Nonetheless, Brighton was both imagined as 'safer than anywhere else in the UK' (Questionnaire 517) and experienced as such by some, 'I feel so comfortable being who I am here' (Questionnaire 602). Many spoke of the perceived dangers of 'elsewheres' and the 'safety' of Brighton. Imaginings of Brighton as 'safe' were formed, paradoxically, not only through feelings of safety but also in the acknowledgement that something might happen in particular parts of the city, but this could be prevented and safety achieved through monitoring behaviour and avoiding 'dangerous' areas. Over the course of the first decade of the twenty-first century, numerous safety initiatives worked to improve the safety of LGBT people by focusing on the reporting of 'hate crime'.[5] Such reports were used to create evidence of a problem, as well as enabling victim support by identifying those who needed it. The problem then was individualised violence (often seen as isolated and therefore punishable) between straight and gay people, and was

---

5   In this context 'hate crime' is differentiated from hate incident. A hate incident, according to the ACPO, is 'Any incident, which may or may not constitute a criminal offence, which is perceived by the victim or any other person, as being motivated by prejudice or hate.' Whereas a hate crime is: 'Any hate incident, which constitutes a criminal offence, perceived by the victim or any other person, as being motivated by prejudice or hate.' (Stonewall 2011). While the Brighton and Hove police monitored hate incidents, they only investigated hate crimes. In this chapter, where we refer to hate crimes with this understanding we use the term 'legally defined hate crime' to indicate this.

not related to systemic discriminations and prejudices and everyday abuses. The solution proposed was through policing:

> Edward: Brighton and Hove is a safe place to live. Now that is proven and backed up by statistics of crime falling, and it's proven by the fact you can walk down St James's Street [where a number of gay businesses are located, see Chapter 1] and you will see a trans lady and you'll see two men with beards and they'll be holding hands. And if it wasn't a safe place to live, you wouldn't achieve that community product.

> The way we [the police] manage a public sex environment is [a] highlight. I say it from the fact the Government Office for the south-east now recognise us formally as being an ambassador for the way we managed this issue in the city. I can further evidence that by the fact the Metropolitan Police, Surrey Police and other divisions in Sussex, have come to us regarding how they should manage their own public sex environments. Brighton's is quite unique really. The way it's managed. The way we police it. The way we deal with the community. And it is very good.

> [Name of Public Sex Environment, PSE] is not policed for a purpose of catching people having sex.[6] It's policed with the purpose of making sure everyone's safe. If a gay person at [name of PSE], like anyone in the city, was a victim of crime in that area, and [name of PSE] would be no different to [name of street], then we would provide a higher visible presence to ensure people felt safe and we caught offenders and protected the vulnerable. So we just apply those same principles to [name of PSE] that we apply everywhere else.

Edward, who worked for the police, spoke of Brighton as a safe place, with his evidence lying in a 'community product' produced through reactive policing.[7] He argued that the supposed safety of marginalised LGBT people ('a trans lady') was

---

6   Challenging the predictions of the 'vanilla-isation' of the LGBT community in response to incorporation into community safety initiatives (Bell and Binnie 2000, McGhee 2003: 366), and the 'desexing' of lesbian and gay initiatives (see Cooper 2006, Monro 2007) these PSE policing procedures (as result of significant work and pressure by community groups) instead revealed a lack of compromise or 'straightening up' public sex environments. Ordinary was redefined here.

7   The notion of a 'one size fits all' anti-crime initiative and treating sensitive spaces 'like everywhere else' neglects the histories of policing that continued to be salient for LGBT people. This is a reactive understandings of 'safety', that relies on a dubious and contested view that police presence in public sex environments equates to feelings of safety (indeed it might be the danger that is arousing). Moreover, as Moran (2007) notes a small minority of incidents occur in these areas, and of the 598 people in CMIT who experienced some form of hate crime, only 27 experienced an incident in a cruising area (see Browne and Lim, 2008c). Moran (2007) found a resistance from safety services and LGBT communities

an indicator of safety for the LGBT collective and this was achieved by punishing legally defined hate crime. The 'trickle down' effect of 'catching' (straight) perpetrators and the presumptions that dealing with gay men's sex areas will support trans (and bi) people's safety is a point contested by our data (see Browne and Lim 2008b, Browne and Lim 2010, Browne, Bakshi and Lim 2011). LGBT safety was thus perceived by some as a monolithic item ('a community product'), such that addressing parts (usually violence by straight strangers against gay men in public places, see Moran 2007) supposedly had wider effects that helped to achieve LGBT safety. Brighton's police force was supposedly leading the way and this is recreated by imagining the city as somewhere that *should* be safe for LGBT people.

However, some LGBT people continued to feel unsafe relating this to more than the numbers of occurrences of reported hate crimes. This research found that feeling unsafe related to knowing about *any* violence against other LGBT individuals, as well as from unpleasant interactions (such as 'funny' looks, see Browne 2006a). Although importance was placed on the recognition of abuse and retribution through institutional channels, the claim by the police of falling crime rates were read as a loss of confidence in reporting, rather than a 'successful gain' in policing. Yet, the claim to retribution and criminal justice does not equate to safety in its diverse and multiple manifestations, perhaps because it does not enable the attainment of ordinariness.

Police strategies were explicit about pursuing an agenda focused on legally defined hate crimes, rather than broader safety initiatives that seek social change. Perhaps those who sought to reactively police hate crime relied on the underlying assumption that it could not be prevented and that LGBT people are inherently vulnerable. From this perspective, the aspiration of safety in its broadest sense was unreasonable because it was unattainable.[8] And yet Brighton was seen as somewhere where this ideal was pursued, driving a form of activism that expected better:

> Jack: My job's split between case working [for] LGBT people who experience safety problems: that might include hate crime, it might include domestic violence, it might include sexual assault, all of those. So, that's half of my job, which generally involves supporting people. It involves advocating for them with other organisations. It'll include making sure their housing's sorted out, making sure they're safe where they're living. It will be giving them a safe way in to police processes.

---

to deal with the figures from research undertaken in London that suggested the focus of LGBT safety is better placed elsewhere.

8   In Perry's (2001) terms such a call for safety would alter social orders, making the 'disorder' of LGBT lives part of the social order. It could also question dominant modes of masculinity.

I think the police has a lot of mystique around it, actually what the police do is very simple, it's about taking a report of something, it's about investigating, and then taking it to the Crown Prosecution Service who decide whether they can charge it or not. My job is really all the other stuff around that. Making sure that all those other things that don't get addressed by that – helping to address them. And the reason why it's LGBT-specific is because obviously LGBT people have a lot of un-met needs around their safety. The other things that people might not think about so much is that I deal with disputes between LGBT people, LGBT people getting ripped off, LGBT people getting their drinks spiked, that kind of stuff. So the other end of it is about outward facing stuff, so that will be working with the different organisations in town and the LGBT working groups. It'll be setting up new campaigns. It'll be doing press releases. It'll be strategic work internally, so looking at policy and practice.

Whereas some were concerned with 'punishing criminals', Jack (civilian LGBT liaison contact in the police at the time of interview) was concerned with safety more broadly and augmenting the punitive focus of criminality. Thus working from within the police,[9] the remits of safety could be expanded beyond the bounds of reactive punishment. This worked towards a form of ordinariness through a preventative model that did not assume that LGBT people are necessarily subject to violence and abuse. It required a shift in understanding the possibilities of preventing and reducing violence against LGBT people, rather than dealing with it when it happens.

Despite the desire of LGBT community groups and others to work for ordinariness for LGBT people that meant that they could be free from violence and the fear of abuse, histories of partnership work in safety in Brighton were fraught. Perhaps because of the contested models of what could be considered 'safety', the complexities of ordinary activisms with and within the state were apparent, as we now discuss.

The 'Danger Forum': Contestations of and within Partnership Working

In the UK the move from 'crime prevention' to 'community safety' in the late 1990s placed the onus of coordinating safety in the hands of statutory partners, with the 1998 Crime and Disorder Act requiring community consultation and work across local authorities, the police and other bodies. These 'partnerships' were also required to include a wide range of community and voluntary groups, including lesbian and gay groups that were read as 'hard to reach' (McGhee 2003: 350,

---

9   The diversity of 'roles' and conceptualisations of 'safety' within 'the police' as well as the enactment of these and other policies illustrates that 'the police' are heterogeneous institutions. Similar to 'the state' this was diversely enacted through assemblages of aspects such as legislation, discourses and individual as well as collective actions.

2004). Taking a holistic view of hate crime challenges a single agency approach to solving crime, where the police are seen as 'the' agency (Crawford 2001 in McGhee 2003). More than this, the inclusion agenda created the imperative to 'invite the full participation of gay and lesbian groups in the work of the new partnerships' (Home Office 1998: 2.44, cited in McGhee 2004: 365). We now explore the attempts to create safety through partnership working enacted in structures such as the police, the partnership community safety team (PCST) and the gay media (see also Moran et al. 2004, Moran 2007). This exploration is set within a national context where policing was increasingly 'partnership led' with the police seeking to work with, rather than 'against' LGBT communities in an equalities era (see McGhee 2003, Moran 2007, Yarwood 2007). Yet, despite success stories in other part of the country (Cooper and Monro 2003, McGhee 2003), LGBT safety partnerships in Brighton were hampered by histories of failed community engagements. Here we examine the complexities of partnership working by outlining the demise of the Brighton Lesbian and Gay Community Safety Forum (and part of the Anti-Violence Initiative, AVI and the Anti-Violence Unit, AVU), set up in 1998 'as a multi-agency response to the problem' of mistrust in, and fear of, the police (Ourstory 2001).

Partnerships did not always follow neat partnership ideals. On the one hand, community partners may not be amenable and can become oppositional and disruptive and block and stall key initiatives. Indeed, partnerships inevitably have conflict and tension (see Craig and Taylor 2002, Tett 2005, Newman and Clarke 2009, Monro 2007). As Newman and Clarke (2009: 61) argue, 'ordinary people … may prove sceptical, doubtful, calculating or compliant or even … irritating[ly] innovative', and when they are 'summoned' communities can be 'reluctant to materialise' and present as 'plural, contradictory and contentious entities' with intersectional differences. On the other hand, partnership working, advocated to empower LGBT people, also had the potential to reiterate existing power relations between 'community' and 'local government' (see Anthana et al. 2002, Craig and Taylor 2002, Newman and Clarke 2008):[10]

> Joseph: There was a golden period of about 18 months when everybody was working together and then we got the 1.2 million pounds for the Anti-Victimisation Unit. Then it all went wrong. The police and the council ended up not valuing the contribution of the community partners because they thought

---

10   We explore this history by quoting Joseph and others at length. Recognising that all narratives and recollections are partial and reconstituted, we use participants' words to illustrate how this story was told. Its importance lies in its continued significance, rather than disputes that arise with regards to 'accuracy'. Of course such disputes would have to hold that there is a 'truth' to be found to be retold, something that has been roundly critiqued for decades in progressive historical thinking. (There has, however, been a backlash to such poststructural engagements and a reinvigoration of the notion that there is 'a truth' that can be 'known'.)

they knew best on everything. Then we found out, especially the council were just being dishonest with the money. I was able to provide the proof that the council were then putting in expenses to the budget, which had nothing to do with the budget. So then the community money starts going down and the money is going to the council. The council were getting this fabulous project for nothing and all the community people were doing the work. Half way through [a three year project] it the independent Home Office evaluators produced an interim report, which highlighted problems with the management of the council.

Under the 1999 Local Government Act, local authorities were supposed to lead multi-agency crime and disorder partnerships. Barriers to the ethos of redistributing power included short timescales and funding-led partnerships that reiterated the power and knowledge networks within local government (Craig and Taylor 2002). In Brighton, success in securing funding to create the AVI was followed by a disintegration of the values that led to the funding being awarded. The (difficult and lengthy) processes that were needed on both sides to address years of antagonism and prejudices were not undertaken. Instead CMIT participants spoke of the erosion of respect and empowerment, as well as a questioning of the financial management of safety partnerships. Powell and Dowling (2006) note that there was the potential for coercion in these unequal partnerships, despite the ethos of redistributing power. Yet, in Joseph's narrative we can also see activist resistances, questionings and challenges that illustrate agency, as well as activisms from within. Indeed Joseph saw himself as responsible for closing the forum:

> Joseph: I closed the Safety Forum down because I saw it as a danger forum rather than a safety forum. If people couldn't come to a safety forum and know they were gonna be safe and be treated with respect, you can't have one.

> Interviewer: Why couldn't people come?

> Joseph: The council, because the trans people were just exposed. One of the people on the safety forum was a transphobe, who was on the interview panels for the jobs. We recorded him talking to a trans person on the telephone calling, calling the girl vermin and trash, and the people that lived in his house were threatening to beat her up in the background. We took it to the police. The police were mortified. [They] were great 'cause they knew then they had a big problem. The council wouldn't accept it. They [were] happy to work with this person and happy to have him interview [woman's name] who was then transitioning from [man's name] and that was the issue. So we just closed everything down.

> Then [because of] the '98 legislation – they have to have a forum to work with. That's the problem. At first they tried to start their own one and invited everybody to the police station. They made a big mistake. Everybody that went to the police station understood that they were being used as dummies to try and

create the impression that the police were consulting with an independent forum of their choice that they had invited to the police station. They all went for the first meeting and everybody walked away except for one person [man's name 1] who then was appointed as the Chief's Constable's LGBT advisor and has kept the role ever since.

McGhee (2003: 362) notes that in Southampton's work on partnership community safety, the LGBT community were represented by 'highly motivated and extremely well organised gay men from local health promotion agencies and HIV services'. Joseph, as a powerful player in the city, closed down the safety forum once his trust was gone, leaving a gap in the fulfilment of statutory obligations of the police. The attempt at consultation without engaging with the fall-out of the Lesbian and Gay Safety Forum resulted in one person standing in for 'the community'. This was not unusual, and Fuller and Gedes (2008: 274) discuss how 'radical activists' resigned from local strategic partnerships in the English Midlands to be replaced by 'passive community representatives'. In pointing to the importance of 'gay friendliness' in creating safety initiatives, Joseph moved beyond a punishment model, to instead promote a broader remit of safety that does not simply pick up the pieces of the 'gays being beaten up'. The lesbian and gay safety forum was set up to address the problems of homophobic reactions from the police and to improve reporting (and in this way address community safety). Yet, transphobia and homophobia in the actions of officers and council managers worked against the possibilities of safety for all LGBT people beyond criminal justice and retributive models, reiterating abuse from within institutions who were supposed to be former enemies.

While the police were able to 'tick the box' of community engagement, the breakdown of the Lesbian and Gay Safety Forum caused an extensive rift between and within community groups, the gay media and statutory safety services. This led to particular forms of activisms, characterised by opposing 'them' (council and police) and 'us' (gay and lesbian community). Yet even in contesting and challenging the state, the legal requirements to work with communities meant that there was a negotiation of relations that polarised them and us:

Interviewer: What was the role [of name of media outlet] then?

Joseph: We had a public voice to challenge them and they were terrified of it. I put the quotes from the AVI report on the front cover, which was from serving officers who I'd interviewed for the jobs as being gay friendly, saying 'all these gays deserve to be beaten up'. All of this was in the report they were trying to suppress because they suddenly realised that these people weren't gay friendly at all. I said in the editorial, 'I'm gonna leave this in there every month until the police and the council sort this problem out.'

It took about four days and the Chief Constable came to my office. I shit myself [when] they rang me. First of all his whole team came in and checked out the office, for wiring or whatever. I thought he was gonna come and tell me off, didn't sleep the night before. I got the office to get out all the clips that were anti-police, to put them up on the wall just to remind him that we were on the case. He was great. He said, 'I think you're trying to send me a message', that was his opening line. [Laughs] I said, 'you could say that'. He said 'tell me what's gone on', so I just told him everything. He said, 'I've taken my eye off the ball in Brighton' and 'there's clearly a problem here that needs sorting out'. His pitch was that he needed three years to sort the gay problem out in Brighton. He told me what he was gonna do and I said 'fine'.

The good thing is that as soon as they know the gay press is asking questions, now they react in the same way as they would do if the Argus [local newspaper] were asking questions – they get nervous straightaway.

Resistant modes of working were intermingled with collusion in Joseph's narrative. It might be up to the police to 'sort the gay problem out in Brighton', dealing with systemic homophobia of those involved in policing, but this was done with the gay media 'back on board', illustrating the relational production of the state and non-state, in part through partnership working. Thus, while partnerships can fail, this is not the end rather the requirement for partnership working necessitates negotiations between 'them' and 'us'.

Another fraught matter concerns who comes to represent and speak for 'us' at these negotiations (see Chapter 4). In Brighton, certain individual gay men were placed in powerful positions and liaisons with state actors. These positionings can be effective in 'getting things done', but they are also problematic. While Joseph speaks of 'closing down' the safety forum, he does not associate his positioning with that of the one gay man who became the representative of 'the gay community', despite secret meetings with the police and others. Newman and Clarke (2008) note that community representatives may act in exploitative ways. The period of partnership working in gay Brighton at the turn of the century was characterised by more than 'the gay community' standing up to the perceived failings of partnership working with the police and council. One participant referred to the whole scenario through the metaphor of cannibalism, where gay and lesbian communities 'ate each other'. Many lesbians felt let down and personally vilified through the process, particularly by powerful gay men:

Louise: In the Anti-Victimisation Unit there was a sustained attack on women rather than a sustained attack on the police as an institution that traditionally has a very patriarchal way, an institutional way, of doing things. It was very [much] sustained on individuals rather than non-individuals. I do feel that.

Richardson and Monro (2012) argue that the problems with participation relate to resources and capacity, and that bi and trans people are particularly under resourced, with the loudest LGBT voices being listened to while certain other

LGBT people are not heard. Even where lesbian voices were supposedly part of the conversation, Louise's narrative points to how them/us activisms can be manifest through creating gendered hierarchies within an 'us'. Unilateral actions by individuals were important to understanding many activists' experiences of working within and 'for' LGBT communities. Those who resist the state and other normativities can be lionised and celebrated. However, Louise highlights the gendered ways in which power complicates discussions of them and us in addressing activisms that resist state processes, as well as those that work in collusion with new normativities.

The complex working of safety partnerships were crucial to the (re)formation of safety initiatives and policing in Brighton throughout the early twenty-first century. As we will see in the next section, the city's history reconstituted the forms of LGBT activisms, as well as who was part of the partnerships. We now explore how the legacies of the Lesbian and Gay Safety Forum, the AVU and the AVI informed police and local government's actions in the first decade of the twenty-first century.

Policing Gay Brighton

Yarwood (2007: 447) contends that 'a better understanding of the police contributes to a better understanding of the ways in which power shapes space'. In the years following the demise of the safety forum in 2000, a new civilian post was instigated in the Brighton and Hove police, that of the civilian LGBT Community Liaison Officer. A key part of this role was dealing with the anguish and relationship breakdown discussed above. This post and the person that occupied it altered the ways in which them/us boundaries were created. Creating this LGBT post from within the police renegotiated the terms of inclusion and community engagement:

> Harriet: It was advertised as 'LGBT Community Liaison Officer'. I thought, 'Great, it's going be going out to communities and talking to them' and thought, 'It's Brighton, there's not going to be that many homophobic incidents, it's going to be fine'. I got here and it was just absolutely shocking. At that time there was the massive breakdown between the police and the community and all the front pages of like [name of media outlet] and all the other press were really negative against the police.

> I was really kind of cautious when I joined that role. I didn't want to take political sides. I just engaged with absolutely everyone and just took on all their kind of comments. So I made sure that they all felt that they could actually work with the police, rather than either siding with one or the other. So that was one of the key things, not to actually believe all the stories, but [to] know [that] there's obviously been something tragic happen[ed], to actually get to that point.

Harriet's assumptions regarding the safety of Brighton drew on imaginings of the city that were quickly disrupted. She sought to move beyond polarised politics, looking to encourage working with the police. Her work also questioned and challenged power relations and backroom deals that had characterised community safety initiatives prior to this point. Showing that key individuals can have significant effects on partnership working (see Asthana et al. 2002, Cooper 1994), Harriet extended the representation of the LGBT community away from one gay man, or the gay media seeking to engage a range of stakeholders including those marginalised and disenfranchised from policing processes.

The legacies of safety partnerships of the city reiterated entrenched positions and painful histories of community engagement. These contrasted with Harriet's imaginings of Brighton prior to her moving to the the city and taking up the post. The legacies safety partnerships in the city were enacted through heated and difficult meetings:

> Harriet: One of the first main community meetings that was held after I got into post – I think it was easily probably about 100 people in the hall that turned up. I was really shocked that so many people were interested. One of the key things that came out was that they'd said they weren't happy with the way the police were running things and they weren't actually listening to their needs. The police had taken the only person, which was [name 1 as above] at the time, who was happy to stay engaged and give an opinion. They were like literally riding on his every word and doing all the stuff that he suggested and it was all good intent, but at the same time he's one gay man, he doesn't represent the whole of the LGBT community. That upset people even more, especially within the female part of the gay community. There was a lot of like backlash about that.

> The meeting got really out of hand a couple of times. I remember trying to calm it down and it felt really unproductive, because everyone was shouting and disagreeing with each other. I remember thinking, 'How on earth are we actually going to get through this, because the community can't agree with each other to start with, so how are we actually supposed to engage with this community, when they can't agree?' It was just carnage really.

> From there on I helped set up the LGBT consultation group with the police. We were saying we want as many people as possible to get involved and come along. We did have quite a big group then of different people from different parts of the community that came along, but I'd still say it's only a small representation, really, and not massive as we would have liked. Things just started getting better from there. We started putting a lot more communication out and letting people know what we were doing and asking people to get involved if they had any issues or they wanted to raise anything. Just be more open and transparent about it really. Just going to all the different political groups and all the different

communities and businesses and trying to get them all on board and make sure that they were all being heard, rather than just listening to one.

In contrast to the ways in which diversity between LGBT people can be used to stagnate work (see Chapter 6), Harriet sought ways forward that navigated these differences and the fraught histories and relationships that defined community safety in gay Brighton. This meeting led to sustained work which sought to reverse the impact of the legacy of the Lesbian and Gay Safety Forum, particularly between those designated as LGBT communities and the police.[11] In this way partnership working was reconstituted not only through specific local histories, but also through ongoing imperatives to work with LGBT people.

Despite the safety initiatives and individuals who sought to readdress not only the failure of partnership working, but also long standing LGBT relationships with the police, a lack of confidence persisted:

> Rosa: I had a landlord who punched me physically in the face because I was trans, give me a load of verbal abuse in his shop. When the police arrived I was the one who was then threatened with some sort of order which meant I couldn't go near his shop, on the basis of me being transgendered. Another example where I was living, we had people next door who were bringing in guys at four o'clock in the morning. The guy punched me in the face, we had the police in. I was the one who was made to feel as if I was the freak, because I was trans. Recently, I haven't seen that, in the sense that the police seem to be trying really hard with the LGBT community. The point is, though, that there are no trans coppers, so that would be nice. But I think that the police are actually trying to treat us like a complete community. I think they are trying to treat us as an LGBT community.
>
> Interviewer: After all those experiences, would you be happier reporting if something happened?
>
> Rosa: That's a really good point. No, that I'm quite clear about. I would really like to have reported those things to a transgendered copper. No doubt about

---

11   The work of Harriet and the other LGBT officers in this area reaped some benefit. The CMIT research found that attitudes to the police had changed in the five years leading up to the questionnaire in 2006. Fifty-eight per cent (n. 225) of those who lived in Brighton and Hove for over five years said that the police improved in the five years prior to the research, while 38 per cent (n. 148) were not sure. These figures were supported in the qualitative data. Although the perceived likelihood of getting beaten up was the same as anywhere else, its police service was seen to be making Brighton different in terms of how this abuse was dealt with. Yet, the duty of the police was not to reduce the likelihood of getting beaten up, but rather they would be looking to help LGBT people if we got into some sort of trouble.

that at all, because they would have understood that from me. I would have been more comfortable with that.

Interviewer: Will you report things now though?

Rosa: No. I mean that is a really excellent bloody measure of my real trust in the police. Would I report it now? No, most probably not. I just feel that in some way I would be victimised still. We know there's massive homophobia still within the police force but we do know that they're trying to do something about it. So no, I just don't know that I would, end of story. (Trans focus group)

Rosa's narrative indicates the importance of both being part of an LGBT grouping in order to be 'dealt with', while also wanting the possibility of reporting to trans 'coppers' who would understand her experiences. Ordinary lives, in this context, are achieved through feeling safe, which can perhaps be attained by being simultaneously within homonorms to feel safe, without negating the differences that make trans 'coppers' appealing (see Chapter 9). Police initiatives in Brighton sought to address the wish, articulated by some trans people, to see 'people like me' in the police force. The burden of representing 'a trans community' can negate diversities between the 'us' that is supposed to be catered for. Nonetheless, there were possibilities for activisms that created ordinariness through working as the police, as well as with them in partnerships.

The existence of an LGBT person (who was bi-identified) *within* the police after the Lesbian and Gay Safety Forum fall out had implications for how activisms developed beyond the them/us dichotomy and the acknowledgement of differences between LGBT people. Harriet challenged not only homophobia, identified in earlier eras, but also biphobia and transphobia in the police. This was undertaken in part through inviting in bi and trans people and community groups into police training to tell their stories, offering a different form of activism that sought liberations from within:

Harriet: When we did the LGBT awareness training there was loads of officers, especially your old school that had been in forever [who said] 'I live in Brighton, I don't need any diversity training, what's this LGBT shit?' I just thought, 'This is going to be absolutely horrendous'. The way we did the training, in the afternoon we'd have different people from different areas of the community as volunteers, who would come in and speak about their experiences and we'd break them into groups. After one of the training, there was this guy, really old school and you wouldn't put him down to be anything like supportive towards stuff like this. He's like, 'I did think I don't want to go to this training. But I have to say to you, I really, really enjoyed it. I found it really interesting. I was just amazed to hear the experiences of people.' I think especially the trans stuff, because they heard all these stories and they were like 'Oh my God'. Just realising how serious it is, how it's affected people's lives. It got them to see

> what it's really like for people and I think it did change quite a lot of opinions. We started up a trans-working group as well. That obviously showed them that the police were wanting to change.

> I was really surprised by how supportive the police were and they took it all on board fully. Stuff that I thought probably might not actually go through, did go through and they were all supportive of it.

Harriet's surprise at the 'supportive' stance of the police on trans and other LGBT issues highlighted the continuing perceptions of Brighton police as homophobic, biphobic and transphobic. Yet, Harriet's experiences of diversity training were seen as altering opinions for the benefit of marginalised LGBT people. The focus on one identity category could be critiqued here as failing to recognise the fractured ambiguities of gender and sexual identities that are always inflected with class and race, and of course working with the police does not address the problems of state violence by the police (see Spade 2011). However, as Moran and Sharpe (2004: 411) note, trans experiences can offer one way of 'highlight[ing] the interconnectedness of social hierarchies and the place of violence in their simultaneous reproduction'. Certainly police initiatives have sought to work with and for trans people and were appreciated by some trans activists, who understood the police in Brighton as flawed but 'trying'.

Place was once again highlighted as crucial to these ordinary activisms and illustrates how imaginings of place can operate in diverse ways. While in Chapter 4 Brighton's LGBT reputation was seen as an impetus to undertake work, Chapter 6 pointed to 'resting on the laurels' of the supposed 'success' of LGBT equalities in the city. Here place-based assumptions presumed that tolerance was accessed simply through being resident in the city, and this paradoxically was experienced as working against training that sought to explore the difficulties of LGBT lives in the city.

Although working for the police enabled Harriet to enact activisms that sought to make LGBT lives ordinary from within, these continued to be fraught in the broader context of 'safety services' in the city. For some the safety services provided by the police and the council were experienced as 'perpetrators' (Questionnaire 693) who have 'never really been in an effective state' (Leonie, individual interview). Spade (2011) questions benign and neutral understandings of 'the state' and institutions such as the police, arguing that these are formed through race, classed and other social differences in ways that perpetuate a particular social order. In addition to these key social differences, place is an important constituent of social life. For example, Leonie argues that the 'mess' of Brighton stifled partnership possibilities, emphasising the diverse ways in which partnership working (and thus the state and the police) were manifest in gay Brighton:

> Leonie: I think for LGBT people the equalities agenda has been embraced across many areas, but [the] way it's been embraced in hate crime has been particularly

ineffective. I don't understand what it is about hate crime that makes it so difficult, but hate crime provision is in such a mess. I think the statutory services dealing with hate crime really have, I wouldn't say lost the plot, but I don't think they ever really got the plot, it's never, never really been in an effective state.

Leonie points to how safety 'partnerships' were broader than police–community relations that have been the focus here (see also McGhee 2003). The ineffectiveness and 'messiness' of 'embracing hate crime' needs to also account for the Partnership Community Safety Team (PCST) based in the local authority of Brighton and Hove city council. The next section explores the PCST illustrating how activisms that sought to make LGBT people ordinary operated through resistances, as well as cooperation.

The Partnership Community Safety Team

The sharing of safety across local authorities as well as with other bodies has arguably reduced the responsibility (and accountability) of the police for safety, instead devolving this responsibility to 'active citizens' (mainly unpaid community volunteers) through partnership working (see Yarwood 2007, McGhee 2003). Thus, when examining safety in Brighton, how the Partnership Community Safety Team (PCST) representing the local government engaged in safety partnerships key. This facet of Brighton and Hove City Council presented itself in this way:

The Community Safety Team works in partnership to improve community safety, reduce crime and disorder and make people feel safer across the city. (Community Safety Team, Brighton and Hove City Council 2011)

The Partnership Community Safety Teams' initiative, mandated by central government, was part of how equalities legislation was enacted at a local government level in Brighton. Reflecting the ethos that safety was key to inclusions, these teams covered areas such as racist crimes and incidents, disability hate incidents and LGBT hate incidents and crimes. In Brighton the specific team dealing with LGBT issues, described its priorities as:

Working hard to provide support to all those who experience incidents motivated by prejudice as well as bringing perpetrators to justice. (Safe in the City 2011)

Continuing with a model of justice and punishment, the Brighton PCST further perpetuated a discourse of safety in a limited way, namely supporting 'those who experience incidents' and bringing 'perpetrators to justice'. In contrast participants in CMIT discussed how local government safety services *should* have acted to protect LGBT people as part of the wider remit to deal not only with crime, but also to promote safety:

Matt: The people that are harassing us, this has been going on for quite a while now, and for some reason the council won't take action. I understand that they might be ill or whatever, but I'm a guy living with HIV. I changed my medication this year and I failed a combination that [had] side effects like you couldn't believe. I couldn't believe it myself. I had to stop and in the meantime I'm getting called 'queer' and 'faggot' in the street and I thought 'I don't need this'. The council are there to house people, some vulnerable people for different reasons, but surely they should be there to protect me. The council have failed me as a landlord, they really have and I'm disgusted with them. (Hate crime focus group)

Questioning the stranger danger model that continues to dominate discussions of safety (see Moran 2007), Matt discussed ongoing harassment from his neighbours in social housing. He argued that not only did local government services fail to 'take action' and that they should not have simply reacted to the violence he experienced, but also protected him in the first place. This moves beyond a criminalisation model that seeks redress, towards one that asks services, such as the PCST, to work to 'make people feel safer across the city'.

It is clear from Matt's narrative that the local government's authority and responsibilities extended to the home and that failures in the safety remit of the local authority had not only perpetuated the abuse, but also became part of the abuse itself.[12] Even within a retribution model Brighton and Hove City Council (as well as the health service) was trusted less than the police as a key partner in safety initiatives:

Persia: I know that if some police officer did some transphobic thing with me, I would be seriously listened to immediately and immediate action would follow. I know this. But I don't know that in the council and I don't know about the NHS either. Oh they fudge and they poke around and they don't do it. And that's what I'd like to see, super gold standards, zero tolerance. I've mentioned it and they go 'oh it's a bit much to take in', fudge, fudge.

In contrast to the police, whose community relations were praised in CMIT, there were concerns relating to the Brighton PCST's work in the area of LGBT initiatives. There were regular breakdowns in relationships between those supposedly involved in the safety partnerships. This set up them/us binaries that blurred the lines of LGBT and non-LGBT. Relations were fraught and activist 'burn out' was apparent in this area, with numerous LGBT PCST employees 'coming a cropper' (Joseph). In this way we can see that the legacies that created the LGBT section of

---

12   Following the publication of the housing findings, Matt and his then partner were offered alternative accommodation. Of course the question of whether they should 'have to move', was a vexed one that is addressed in Browne and Davis (2008). In this case, following the ongoing abuse Matt and his then ex-partner chose to move.

the PCST were based on resistant, binary and oppositional relationships. This not only illustrates the centrality of local place-based histories, it also highlights how partnerships are reformed through place:

> Joseph: [In 2003] they cobbled up this nonsense, which is called the Partnership Community Safety Team. There was no joined up thinking. It was just like a face-saver for what was going before, no lessons had been learned. I've given up on them [PCST] years ago and I don't think it's possible to do business with them.

> I'm not going to send people to a place where there's [someone incompetent or incapable] giving out advice. I'm just not prepared to do it. I had the conversation with [name of manager] and I told [them] what was going on and [they] said to me, 'well, you've got to get them to make a complaint' and I said, 'no, [name of manager], I don't. You have a duty of care to the victim, sort it out'. That was the same conversation I had with [them] over [name of trans person involved in the lesbian and gay safety forum, see above]. I said, 'you're the manager, you go and ask the questions. If you ask the questions you will get the answers.' The interim evaluation report for the AVU identified [this manager] as a problem. Here we are eight years on with [them] running a unit that nobody knows about.[13]

In contrast to the changes identified in the police service, the LGBT aspect of the PCST was understood as remaining stagnant, and Joseph recounts that the problems identified at the time of the breakdown of the safety forum went unaddressed (to the detriment of staff and of LGBT people who needed support). An individual Brighton and Hove City Council manager was often named as a key aspect of 'the problem'. As Asthana et al. (2002) contend, 'key personalities can act as barriers to effective partnership working' (788). The temporal fragmenting and 'forgetting' practiced by state actors in the PCST meant that, as with other local authorities, it was 'without memory or consciousness of its past' (Cooper and Monro 2003: 241, Cooper 2006). The lack of attention to the problems perceived to be inherent to the PCST illustrated an uneven approach to evaluating community and statutory groups. Community organisations, such as Spectrum (Chapter 6), were deemed to be 'failing' (Andrew) due to a general lack of knowledge regarding its activities. However the Partnership Community Safety Team, which was supposedly 'outwardly facing', was not held to account against the same measures. Joseph's withdrawal by 'giving up on them' can be seen as a form of activism in situations where legislation requires partnership working. It also recreates a 'them' that are failing, refusing to allow a blurring of boundaries between a 'them' and an 'us'.

---

13   In the quantitative data gathered in 2006, over half of all respondents were aware of the Police LGBT Community Liaison Officer, but less than a fifth knew of the Partnership Community Safety Team.

Not only was the PCST a unit that 'nobody knows about, that nobody reports to', it was also seen as 'pulling down' other services, establishing a 'them' that needed to be challenged and contested by LGBT activists 'outside':

> Joseph: What they [the police] have done now [2010] is that they've thrown all their eggs in one basket. So where we are now is that we've got a unit, which is in my view dysfunctional. It's dysfunctional because 1) people have problems accessing it; 2) people don't know what it's doing; and 3) [name of LGBT employee], I don't know what [they do]. The police are being told to work in partnership with these initiatives, have got sucked in to supporting the council and that's where the damage is being done. The police have been working I think very hard to try and get things right over the last few years. They are being pulled down by having to support the PCST. I see all the good work the police have done. The workers, the foot soldiers are doing great work and they're being let down by the command team who aren't prepared to stand up to the fact that this unit is dysfunctional.

In contrast to the vision of partnership working, the legally forced relationship between the police and the local authority was seen as detrimental to those who were trying 'to get things right'. Thus, the evenness and diversity 'within' is both comparative (as in the police are seen as doing 'better'), but also relational (as when the police are 'pulled down' by working together). This complexity points to the multiplicities of activisms that operated within as well as against 'failing' safety services that are themselves heterogeneous, multifaceted and diverse.

Placing the blame in 'the command team', Joseph recognises the volume and importance of work undertaken by the 'foot soldiers'. It was not only Joseph who had difficulties in working with the PCST, 'foot soldiers', who can be understood as LGBT activists who are working within, also spoke of the difficulties they had in working with the PCST, indicating firewalls and blockages are not just posed to those 'outside' of the institution. Harriet indicates the complexities of working with *and* against those who you are supposed to be 'in partnership with':

> Interviewer: What was your experience of working with the partnership [community safety team]?
>
> Harriet: Not particularly brilliant. Even though I was part of the team, I didn't get invited to a lot of the meetings for the first couple of years that I was in post. The Race Liaison Officer had the same problems as well, because they haven't got an accessible building and one of the members of staff was a wheelchair user and couldn't get to the meetings. They didn't really do a huge amount about it.
>
> I think there were a few issues internally with LGBT officers not doing any work, not being there half the time and not very good line management supervision. There were a few workers in there that were really good, [who] were also part

of the community and really keen to do stuff and get on board, but at the same time that wasn't actually their role. They were just part of that team as different workers. They did more towards the LGBT community than the person who was supposed to be doing it and yes, it's controversial.

It [PCST] was seen quite negatively by certain parts of the community. Yet that was their main duty to the community, to work with them and do stuff for them. They didn't feel that they were getting the service that they should be getting. Sometimes I was embarrassed to be part of that team and especially when I wasn't getting invited to all the meetings.

I stuck with the police and the community work I was doing and half pretended that that didn't exist. Now and again meet up. Then we started doing the case work meetings. I'd turn up to a case meeting with like 20, 30 cases that I was doing that month and the other person would be like 'Oh well I've got this case and this case' and 'Oh and I've not done this, I've not done that' and it was like 'Am I the only one doing any work here?' Especially when a lot of the cases I was dealing with were not really police matters. They were housing issues and council matters, stuff like that. It should have been the community safety team – who've got the links to the housing associations – taking that on-board. They weren't actually doing it at the time. That was quite frustrating.

I think there's always been a bit of tension anyway, because obviously [name], from Spectrum, didn't have huge faith in the community safety team either. Obviously, he was running a lot of projects and doing a lot of work and they weren't really doing what they were supposed to be doing.

Whereas key personalities can act as 'barriers' to partnership working (see Asthana et al., 2002: 788), they are also conduits to the possibilities of effecting social change from within (Cooper 2006, Cooper and Monro 2003). The frustrations regarding the PCST were shared amongst LGBT groups and the inability to influence this unit can be read as the inability to effectively deploy resources that were supposedly accruing to these groups. Moreover, it shows the limits of activisms from within, where state firewalls/brick walls act as powerful inhibitors to community engagement that they were statutorily supposed to enact. Harriet pointed to the tensions that developed between the PCST and LGBT community groups, through these LGBT activists who can fail to act 'appropriately' under the strictures of partnership working:

Ted: There was a very difficult relationship between that [PCST] team and Spectrum. Spectrum became quite unhappy with what the community safety team were doing around LGBT and quality work. The community safety team were quite unhappy with the way that Spectrum were engaging with them, so basically the community safety team felt that what Spectrum were doing was far

> too aggressive and actually they were worried about the safety of their workers in terms of the kind of abuse they had to put up with. [But] Spectrum had some very legitimate concerns in terms of the kind of work that was happening about community engagements.

> Spectrum and the Partnership Community Safety Team can be seen as having problems with their relationship over the last couple of years. That has been very problematic to a lot of us. It's not the only thing but it's one of the things in the city that's stagnated bits of work happening.

Voluntary and community groups that challenge state actors and fail to act 'appropriately' can be read as 'aggressive' and 'abusive'. Issues between the PCST and Spectrum came to a head when Spectrum formally raised concerns about fitness for purpose of the PCST with reference to individual members of staff. While this action was consistent with accounts from others as above and reflected a view held by people connected with LGBT safety work at the time, this action was deemed inappropriate by management within the PCST (who were seen by some as 'the problem') and stalled partnership working with Spectrum. Raising concerns about the competence of PCST staff was perceived by LGBT activists as advocating on behalf of the community, but impermissible by the PCST who withdrew from partnership work while resolutions were 'pending'. The power to withdraw had very different implications and conceptualisations. Whereas LGBT activists withdrew in order to maintain the integrity of processes and advocate in various ways against power relations that were detrimental to LGBT safety, the PCST's withdrawal reiterated the resources of state actors who could close down partnership working, recuperating the power of the state in these supposed 'equal' alliances. Thus, activisms that work through partnerships can be fraught and tense and a them/us dichotomy can be recreated, such that institutional powers are reformed.

The combination used by some LGBT activists of 'sitting around the table' and simultaneously deploying 'aggressive' (oppositional?) tactics was contested by other LGBT people and activists who sat on 'the other side of the table'. This illustrates the plethora of conflicting ways in which activisms from within can be perceived and enacted:

> Jo: We need to work on it together and not be on either side of a battlefield as it feels that it is like that at the moment, with the community being on one side and the Community Safety Team being on the other. With people like [name of media] in the middle giving food, fodder, and then creating that gap, making it wider and wider still. I think that to make it a safer city, we actually need to work together to do that and we're not at the moment. We're fighting one another.

For Jo, 'fighting one another' fails to achieve desired goals and indeed inhibits the safety of the city. In this way she questions the them/us binary and the view that resistance can be adopted as a potentially productive strategy in this case.

In examining the PCST the legacies of the Lesbian and Gay Safety Forum were seen as being played out, not through backroom deals and changes in policies, but instead through poor management, low outputs and ongoing tensions and relationship breakdowns. The complexity of making Brighton a safer city was apparent not only through contestations and complexities regarding what constitutes safety, but also how LGBT partnership working in the gay capital *should* work beyond opposition. Yet, as we have seen, activisms with and within were diversely, complexly and frustratingly enacted. This contested, as well as recreated, a polarisation of them and us between those who represented the state and those who were placed in the position of community. From this complex and diverse strategies of engagement and opposition worked to make LGBT people ordinary in the sense of being safe in the city.

Conclusion

Ellis: There are serious issues around how community safety and domestic violence and abuse are managed in Brighton and Hove. Although I think there are some really good people on the scene, within statutory services, there's a lack of agreement from the community around what constitutes community safety. There's a wider set of indicators in community safety that is about more than just statutory provision. That kind of stuff that's about wellbeing and community safety meaning not just absence of crime in the same sense that health isn't just about absence of illness. What Spectrum could do and what the [partnership] community safety team could do is start the work that belongs elsewhere, within the community sector, [in] the statutory sector. Keep going, keep plugging away with the statutory sector with all the barriers that that faces, but I would say that because I've already invested in that statutory sector.

Resistant tactics that contest normativities and rely on an opposition between them and us can be celebrated as producing activisms that are better than assimilation. This chapter has shown the complexities of working with and across resistant tactics and assimilation discourses, where neither on their own encompasses the LGBT activisms in Brighton in the first decade of the twenty-first century. LGBT activists who were invited into decision-making positions resisted state actors. Conversely, firewalls/brick walls that prevented LGBT equalities work (Ahmed 2012, Cooper 1994) were used by state representatives to block and stall LGBT initiatives. Brighton was not incidental or background to these processes. Rather, because Brighton *should be* better, it informed how policing and community engagements were undertaken, as well as disappointment in the perceived failings. More than this, the continued 'failures' that were narrated in relation to the legacy of the Lesbian and Gay Safety Forum illustrate how the history of a city builds its present.

While, as Chapter 6 shows, working with the state was operationalised in many areas of LGBT work in Brighton and secured buy in from activists, this was uneven, illustrating the messiness of state enactments (see Painter 2006). Resistances that result in and from opposition did not disappear as LGBT activists were to invited work for inclusion from within. Ellis reminds us that it is not only or simply disputes on how to work through safety initiatives that cause divisions, oppositions, tensions, breakdowns and fallouts between community activists, gay media and statutory providers. Rather the tensions of such relationships rely on (assumed shared) perceptions of what LGBT safety is and how it should be achieved. Safety here is conceptualised as multifactorial in origin such that improving one aspect (such as reporting or policing public sex areas) does not necessarily enhance all LGBT people's experiences of safety. LGBT activist work in Brighton was about more than statutory provision and sought to work beyond (but not neglect) models of safety that were solely located in retribution and 'after the hate crime'.[14] Yet, there was an inherent tension of working across sectors to achieve LGBT safety. A safety model that relies on a post-incident, criminalisation perspective sits uneasily with a model that seeks broader understandings of safety and seeks to move beyond biphobia, transphobia and homophobia, not just react to their manifestations.[15]

We conclude with a document on LGBT safety presented in 2009. This was written after (another) breakdown in the relationship between Spectrum and the PCST. This document follows from the key arguments made in this chapter and was drafted with Leela, in her capacity as a Spectrum trustee. It sought to offer mechanisms to undertake the difficult work of redressing the mistakes of the past and considering safe futures. We believe that its principles offer insights into hopes for change that refuse the placing of LGBT people as victims and illustrate the multiplicities of activisms that are created through engagements as well as resistances. Following this, in Chapter 8 we further discuss the possibilities and pitfalls of ordinary activisms through a discussion of Pride in Brighton and Hove.

---

14   As Perry (2001) contends, broader societal hierarchies that reiterate masculinist (and racist) hegemonies need to be contested in order to effectively address hate crime and improve safety for LGBT people.

15   Moving beyond homophobia, biphobia and transphobia is seductive, but neglects freedom of thought and speech. Perhaps the question for activists and those who seek to work towards LGBT safety then is how can safety for LGBT people be achieved in the presence of these prejudices?

Towards a new LGBT Safety Forum

Renewed community leadership is now needed to steer and progress the broad range of strategic work needed to be undertaken in relation to LGBT community safety.

Functions of a potential new independent **LGBT Safety Forum**:

- **Community-owned action** prevention work, community cohesion, understanding impacts of 'unsafe-ty'.
- **Community engagement** dialogue within and ownership by communities of agenda for change through involvement, consultation open events and fora.
- **Supporting partnership and strategic work** both with LGBT communities and between sectors.
- **Community watchdog** safeguarding the interests and needs of the community and scrutinising implementation of plans [e.g. annual strategic assessment, CDRP strategies].
- **Inclusion of a range of identities and stakeholders within the communities** businesses, voluntary and community sector groups, identity groups (e.g. bi and trans), activists.
- **Evidence-based practice** ensuring actions are supported by evidence, monitoring and analysis of data and evaluation of effectiveness, and supporting further research where needed.
- **Providing infrastructure** resourcing and capacity to undertake above functions.

Spectrum's view is that the above functions are not compatible with one agency acting alone on behalf of the LGBT communities.

- **Lessons need to be learned** from current and previous models here and elsewhere. What are the barriers to increased community capacity and commitment to community safety work?
- **Legacy of disengagement** by and **attrition** of skilled community activists needs to be addressed.
- **Mechanisms for inspiring and nurturing leadership** need to be created.
- **Investment in community capacity** and **delivery of programme of effective action** are key to community sign up.
- **Process needs to be sustainable** and needs commitment from LGBT and mainstream 'leaders'.
- **Plans need to** be designed and delivered to **meet evidenced needs and build community confidence**.
- **Consultation needs to take place with existing working groups** (e.g. LGBT Domestic Violence and Abuse Working Group, LGBT Anti-bullying Working Group, LGBT Housing and Support Working Group) re any proposals for a

new forum as to whether wider community safety issues are to be covered by its remit.

- **Community Dialogue re 'what it means to feel safe'.** Safety ≠ policing. Need to address culture of discrimination and exclusion, impacts of fear of crime and avoidance strategies.
- **Resources and structures for statutory engagement.** How do resources and structures available to support work around LGBT community safety compare to those for racial harassment and domestic violence?

Figure 7.1 Spectrum's Community Safety Document – Summer 2009

Chapter 8

Is Pride Political?: Beyond (Oppositional) Politics in Lesbian, Gay, Bisexual and Trans Festivals

Simon: I did like the city centre being taken over. Last year I was really moved by seeing the Pavilion lit up in pink, because it's been a long time coming that the council's actually embraced it. They're only doing it because it's a cash cow for them but, the fact [is] that they embraced [it] for possibly the first time. I remember [when] they wouldn't allow rainbow flags to be at the town hall.

Frank: It was only eight years ago and I don't think that the councils have moved on much. I think the change in [supporting] Pride is simply because they realise it's actually bringing a lot of money into the city, 150,000 people down, all the guest houses full, pubs full, so suddenly they're embracing it as a tourist thing, not for pride or for the gay community.

David: So why has Pride stopped being something political? Instead it is about something where people want to be in a park drinking alcohol with 120,000 other people, getting out of their heads together en masse.

Frank: I just find it all a bit too fucking negative sometimes. We've all had challenging lives and stuff, but there's been a huge amount of progress in the city and Pride is the example of that. There are 150,000 people who are going to descend on Brighton in two weeks' time. The Pavilion will be pink; everyone will be up for it. I think is a fantastic thing. From those early days when there were a few hundred of us dodging missiles around Churchill Square and trying to chant, 'We're here, we're queer, we're not going shopping' and trying to miss the bullets.

– (General focus group one)

Introduction

Pride parades, rallies and after-parties create distinctive spaces that are significant to understanding contemporary LGBT lives and activisms. It has long been argued that Gay (or LGBT) Pride/Mardi Gras, and particularly Pride parades and marches, are sites of carnivalesque transgressions, where normatively heterosexual streets are re-performed (Bell and Valentine 1995, Gorman-Murray 2009). The focus on heteronormativity and heteropatriarchy and associated disruptions of these norms through parade events have been discussed through the questioning of straight/

gay binaries (Waitt and Markwell 2004, Waitt and Gorman Murray 2008, Johnston 2001, 2005). Academic discussions point to the importance of Global North Pride events in questioning heteronormativity (Waitt and Gorman-Murray 2009, Waitt and Markwell 2004) and the place of consumption in resisting heteronormativity (Kates and Belk 2001). More recently, discussions of homonormativity have placed Global North Prides as corporatised homonormative events, reiterating rather than contesting capitalisms (see Brown 2007, Hughes 2006). As such Prides are conceptualised as re-forming normalities, rather than critiquing them, where LGBT identities are in some way accepted as part of urbanities.

There are distinct spatial differences in terms of the rise of various Pride events and their manifestations in the first decade of the twenty-first century. Prides are diversely structured, although 'the event' may look similar, conforming to the format of a parade followed by an 'after party'. Here, in rethinking Pride events through a discussion of ordinary activism, we are interested in how the location where they occur matters in reconstituting politics. Moving beyond viewing Pride as a one day event to examine the organisation of a range of events that came under the banner of Pride in Brighton and Hove, we explore the power laden and often gendered creation of this.[1] In this chapter we consider the (flawed and imperfect) possibilities of the celebratory politics of pride, that can create ordinariness in spatial and temporally specific ways, without negating nostalgia for an earlier era when political 'sides' were clearly divided.

We open the chapter by discussing the accusation that in 2010, Pride in Brighton and Hove was 'losing its way', contrasted with previous eras. Considering this leads to challenge the notion of protest as the only form of political action. We argue that as a consequence of legislative equalities which altered the position of LGBT people, Pride in Brighton and Hove as an organisation sought to provide space for LGBT people to represent themselves by refusing singular homonormative agendas. We discuss the limitations of these attempts at inclusion, including commercialisations and 'selling out', before exploring the internal politics of Brighton Pride. Focusing on gendered hierarchies and control in gay Brighton, we examine the ways in which lesbians and women can be targeted when in positions of control in the gay capital of the UK.

'Lost its Way': Ordinariness is Not Political

Debate surrounding the politics of Pride events in the first decade of the twenty-first century placed the here and now in opposition with a (romanticised) there and

---

1   Pride in Brighton and Hove was the official name of the event and Brighton Pride was also used more colloquially. We use both interchangeably to discuss the event and when we refer to the organisation that runs Pride in Brighton and Hove, we identify this focus. At times participants refer to 'Pride' – meaning Pride in Brighton and Hove/Brighton Pride.

then. For example, Hughes (2006: 238) argues that 'many gay and lesbian festivals originated as protest marches for gay rights but now appear to have lost political focus and have become more celebratory and commercialised and significant tourist attractions'. The necessary oppositions between party and politics, and politics and commercialism, have been contested in relation to Sydney Mardi Gras, Pride in Brighton and Hove, Dublin Pride and Daylesford Chill-Out festival (Browne 2007a, Kates and Belk 2001, Waitt and Gorman-Murray 2009). The binary of party OR politics however is often retained with an understanding that parties cannot be political. Where politics are read as resisting an enemy, the memorialisation of past gatherings where LGBT people united to oppose an enemy stands in contrast to a diverse and large party in 'apathetic' Brighton where everything is supposedly 'sorted' (see Chapter 3). The loss of the politics of Pride could be seen as a cost of inclusion.

In our data Brighton Pride events in the past were characterised by narratives of a them/us binary, where an aggressor challenged lesbian and gay rights, creating a community that resisted and supported each other. Pride in Brighton and Hove began in its current form with a march/parade and an after event in 1992, protesting against Section 28. These Pride events can be remembered fondly in spite, or perhaps because, of the antagonism experienced:

Geraldine: [In] Pride '92, part of the Pride festival was Queer on the Pier. Lots of us were going to go on it [the pier] and were all wearing t-shirts that celebrated our lesbian and gay sexuality. There were people who had specifically come down from London or wherever to be aggressive and make trouble. In those days [we] used to wear 'Lesbian and Gay Rights Now!' t-shirts with the words as big as possible and walked very confidently in them. On that afternoon, walking along the seafront, several times [we] got accosted by young men. I wasn't hit although I know somebody who was beaten up on that day as a result of that action. But definitely the most aggression that I probably ever experienced in my life happened on that afternoon. And I'm glad we did it. (Women's focus group)

Louise: When they started, it was really small, it was really community and it was really full of LGBT people. What struck me most was the way that the Brighton crowds were watching the event in quite a positive way. Having come from living for ten years in Wales you wouldn't really expect that to happen. It was a lot more political I think at that point than it is now. By the word 'political' I mean there were a lot more floats and people thinking through causes at that time and saying 'this is what we need for L and G people'. So it felt a lot more like the kind of solidarity and wanting change to happen.

The excitement experienced at these events and the desire for more politicisation was framed through these narratives of creating community and solidarity. Enemies made the Pride events political, and singular messages assumed homogeneity ('Lesbian and Gay Rights now', 'this is what we need for L and G people'), against common (homophobic) adversaries (see Faruk, Chapter 6). Such unified causes, coupled with understanding community as something that is small, contrasted with large-scale Pride events.

In the first decade of the twenty-first century, Brighton Pride was transformed by growth in attendance, through both attracting non-LGBT people and tourism associated with Pride in Brighton and Hove.[2] By 2010, Brighton Pride the event was attracting over 150,000 people and consisted of a week of community and business events, a parade and park event on the Saturday and a street party in St. James's Street (see Figure 1.1) over the weekend. While the main event that attracts tourists is in the summer, from the early part of the decade, Brighton Pride also ran a series of events throughout the year, as well as being supported by others who organised fundraising events. These included special nights in pubs and clubs, an annual dog show, feedback events, workshops[3] and Winter Pride.[4]

Throughout Pride week rainbow flags flew across the city, including on all local government buildings. This demonstrated support from individuals, as well as businesses and institutions. In 2004, as Simon notes above, the council paid for pink lights to shine on the iconic landmark of the pavilion (see front cover), an apology for not flying the rainbow flag in the previous years. Such inclusions were reflected in attendances. In 2004 a large-scale survey took place during the park event giving indicators of experiences of those who attended (n. 7,210). This found that a third of the sample defined themselves as heterosexual (n. 1977), the majority of respondents enjoyed at least some of the day (96 per cent, n. 6904) and 80 per cent (n. 3846) of those who have been to other Prides said that Pride in Brighton and Hove is better than other Prides (see Browne et al. 2005 for full details of this research).

During the first decade of the twenty-first century, Pride in Brighton and Hove adopted themes for the summer parade event which were seen by some as frivolous, and at times offensive, rather than causes which could be 'rallied behind':

> Naomi: I miss Pride as a protest rather than a celebration. I remember being
> really excited about Pride in Brighton and Hove when I was younger and lived

---

2 Consequent debates revisited a common argument regarding becoming ordinary, which suggests that if LGBT people were supposedly fully integrated there would be no need for segregated LGBT specific spaces or events such as Pride. In Chapter 5 we argued that ordinariness is not necessarily about sameness with heteronormative values and structures, nor an absence of gay/LGBT spaces.

3 For example, to help local groups with press releases or to encourage intra-community learning regarding Pride week events, putting on a float in the parade and/or getting involved in the park.

4 In 2004, recognising that the summer festival had grown exponentially and had become nationally and internationally significant but was still in need of a local grounding, Brighton Pride established Winter Pride, consisting of a week-long set of activities organised by business and community groups, and which raised funding and awareness of LGBT issues. It was also a commercial 'selling point' for businesses.

in a community where that stuff didn't happen. I think it's lost its way a bit. What the hell is 'Pride Beside the Seaside'? And then the one 'Carry On Pride'. Excuse me? Fucking Carry On films. What year is this? Carry On films are sexist and racist and homophobic, I am not remotely interested in the idea of Carry On Pride. Mostly I just don't feel that much investment in Pride. If it was actually – if Pride said 'Pride against LGBT Bullying in Schools', I would be there with knobs on, I would certainly march in that parade, but it's not.

Naomi points to the exclusions that celebrations such as 'Carry On Pride' can hold, and the lack of politics associated with 'Pride Beside the Seaside'. Many felt a need for a coherent message that works against something, such as LGBT bullying in schools, and unites LGBT people around something that can be opposed and that is specific to LGBT people. This collective endeavour was also disrupted by the inclusion of non-LGBT people, for some, de-gaying the event (see Casey 2004, Skeggs 1999). Indeed the Park and parade events were targeted at this grouping with the aim of educating against homophobia, biphobia and transphobia.

It was believed that there were positive as well as negative effects of the de-gaying of Pride in terms of an absence of transgression, as LGBT people supposedly became normal (at least in terms of the visible spectacle[5]):

Jo: When you kind of look at Pride, the first Pride was over Section 28 and it was a real political march. Then that's turned into a very gender and sexuality neutral event now. My parents have lived here for about four years now and my mum's going to Pride for the first time ever and she's 78. That's because suddenly there's a comfort level for her and an acceptance level for her around things cos it's so normalised. That's good for us because we live here and it's great for people that want to come here. It's 'come to Brighton, this is what you get, Pride'. It's a massive, commercial enterprise. But I don't see Brighton Pride being about the Brighton gay community anymore.

Although Jo suggests Brighton Pride is not about the 'gay community anymore', she still sees the significance of her mother attending the event due to an 'acceptance level'. The normalisation of Pride for straight people, however, meant that it was 'not about the Brighton gay community anymore'.

In the wider literature, the depoliticisation of Pride events has been attributed to the sale of such events as tourist attractions, which can attract both heterosexual dominance and revulsion (Johnston 2001, 2005, Pritchard et al. 1998). Rushbrook (2002) argues that gay spectacles attract tourism, while dangerous or risky spaces

---

5   This 'normalisation' was contested in the pages of the local newspaper that bemoaned the excesses of Pride events, nudity and the lack of 'family friendliness'. These commentaries would emerge regularly when any story regarding LGBT people was reported.

do not. As Jo indicates, ordinariness can be read as lacking transgression and otherness. This was understood as having effects beyond the event itself:

> Gordon: I always blame Pride in Brighton and Hove for the community here not being political because I've always felt that right from the start Pride was more about partying than about message and I've always thought that if we had a more political Pride it would have politicised people locally.

For Gordon, Pride in Brighton and Hove created a lack of politics in the city, and the organisation should politicise and energise people around agendas that 'the community' should support. In contrast Nigella, argued that Brighton itself created a party focus for politics which was reiterated through Pride in Brighton and Hove, ('people [in Brighton] are like "Hey, let's change the world tomorrow, let's have a party, that'll change the world"'). Seeking to take politics seriously ran counter to the celebratory rhetorics of the Brighton bubble (Chapter 3), but fed into and from the idea of Brighton as 'sorted' (Chapter 6). For some Brighton was seen as apolitical in contrast to what 'you do get in other places', which had a better understanding of what 'you are up against' (Nigella). In this way, Brighton was perceived as creating apolitical LGBT groupings who partied rather than fought, because of the perception that LGBT people were part of the city.

For others opposition and fighting against something would not make Brighton better:

> Mabel: Pride is still Pride. That's not a very cohesive thing. That's a reactive thing. We're proud against you who are trying to destroy our identity. That's still a negative reaction. So what positive things do we do that hold us together and celebrate our identity? What do LGBT [people] have in common? Virtually nothing, other than the experience of homophobia, biphobia, transphobia [and] persecution. That's not true of Jews. There are all these positive things that we do. The experience of persecution is not a very positive life-affirming thing.

Mabel noted the links between Pride and shame, as many authors have discussed (see for example Munt 2007, Probyn 1998, 2005). While these authors argue for the productive possibilities of shame, for Mabel, there is 'virtually nothing' beyond persecution that LGBT people 'have in common'. Perhaps this was why Naomi and others focus on the oppositional to define an agenda for Pride, as well as creating community through such agendas. However, as Mabel notes the experience of persecution is 'not a very positive life-affirming thing' in contrast to her view of the rituals and practices that Jews could share and enjoy. She asserts that the identities are built around experiences of persecution and 'positive things that we do', thus opening up the possibility for LGBT people to do other things when coming together for Pride events. A celebratory coming together could sustain the cohesiveness of the category LGBT, through what those who identify with this category do together. Moreover, where Pride was a celebration rather than being against those who 'seek to destroy' us, it could offer positive possibilities, as well as the failings identified here. Holding the tension between the nostalgia for Brighton Prides in the past and the possibilities of celebration, we now move to discuss understandings of the Pride event in the early twenty-first century and ideals of those who organised it. This seeks to explore the potential ordinary activisms offered by Pride in Brighton and Hove in the first decade of the twenty-first century.

Beyond the Barricades? Celebrating Lesbian, Gay, Bisexual and Trans Lives

Questioning the presumed heterosexuality of public spaces can offer a politics that is about 'life affirming moments' for LGBT people themselves (Kates and Belk 2002). This can be political and empowering, but not necessarily in coherent, resistant or radical ways. It is possible to see celebration as political, and being happy as a political enterprise which all are invited to join without conforming to particular normalisations. This can be undertaken without negating failings, limitations and exclusions that all forms of activisms are subject to. We use this section to discuss the possibilities of Pride politics and the ordinary activisms that were created through the event. These included support and celebration.

As we have seen, Pride events were imagined as coming from a context where there were overarching singular identities that united an (imagined) LGBT community. Brighton Pride organisers saw themselves as extending the remit of the lesbian and gay causes identified above and building on the work of those who have gone before:

> Robert (Chair of Pride in Brighton and Hove at the time of the interview): I'm aware, talking to much older gay men, that they used to do the political rallies. The early Prides in the '70s and '80s and things. They got into fights, they were arrested and all sorts of dreadful appalling things, fighting for their rights as a human being and as gay men. Our roots are firmly with them. However, we don't have to fight in quite the same way that they did. The thing that we're celebrating now is a much wider group because from the small acorn that they started fighting for, because they fought so well, it's allowed us to extend that to include lesbians, bisexuals, transgenders, Qs, Us, whatever else, and it gets bigger and bigger. That's a fantastic thing, but sometimes the people who fought feel that they've been forgotten and abandoned. There's a concept that if they can regain the 'gay Pride' label, I really am generalising, they won't be forgotten. We'd be nothing without them. That's fundamentally where all of this stems from, the gay rights movement. But it isn't just gay rights now, it's all those marginalised on behalf of their gender or sexuality that we're fighting for.

Accepting the feeling of the loss of control, forgetting and abandonment, Robert's narrative reflects how the histories of Brighton can be seen as creating the possibilities of Pride in Brighton and Hove. Yet, challenging those who want to 'take it back', Robert argues that Pride came to be about 'all those marginalised [by] their sexual and gender identities', including those marginalised by other LGBT people. Getting 'bigger' in this sense was not about dilution, but instead enabled inclusion of the diversity of LGBT people. Diverse traditions are important to the development and contestation of Pride, including feminisms, black politics and lesbian politics. Recognising these alongside the unified agendas purported by gay men (and at times lesbians) challenges the coherent narratives of past Pride events.

Binnie (2004) notes that visits from 'heteroland' can be empowering for those who attend Pride events and that these can act to bolster, explore and celebrate their gay identity. In the 2004 Pride survey, LGB[6] visitors, slightly more than LGB residents (80 per cent, n. 2944, to 72 per cent n. 706 respectively), strongly agreed that Pride provided an opportunity to express their sexuality more than daily life (see Browne et al. 2005). This supports the ideal of the Brighton bubble as 'better than' other places (Chapter 3). However, it concords with the CMIT research where the majority of LGBT people who lived in Brighton also experienced 'being freer' at Pride:

> Interviewer: What do you think could be changed to make Brighton better?
>
> Paul: More sort of events like gay Pride.
>
> Sharon: Instead of just having like a gay Pride, there should be like loads more gay Prides, loads more, loads more.
>
> Paul: Cos you get one day out of the whole year to announce like who you are. Straight people sort of have their own Pride, like the other 300 and whatever it is days. We have one. (Young people's focus group) [7]

Brighton Pride can feel supportive for those who have recently 'burst' out of the closet, inspiring 'visceral feelings and strong emotions that challenge, counter and often replace the fear and shame experienced' (Kates and Belk 2002: 411). Dancing on a float, experiencing the wonder of being part of Brighton Pride, or tentatively watching from the street, can be 'life affirming', even life changing. In this way Brighton Pride was empowering through the presence of a collective that

---

6   This is used deliberately as it was not possible to include the figures for trans people in reporting this figure, due to the separation of the sexuality and gender question and the inclusion of non-LGBT people in the questionnaire.

7   It is significant that this quote came from a younger people's focus group. In the CMIT research, the first Pride event that was attended by the respondent were remembered fondly, with most going on to talk about 'out-growing Pride' or wanting something more than what they got the first time they attended. This was particularly apparent where they had migrated from places 'where stuff didn't happen' (Naomi, see also quoted above).

celebrated city streets. Happy masses of LGBT people and their allies for some created feelings of belonging, recognition and (temporary) ordinariness, among a wider collective of LGBT people in the city. This was alongside the broader statements that Pride parades can offer regarding the experiences of LGBT lives that are not only, or always, about suffering and rejection.

Kates and Belk (2002) also point to how Pride celebrations can be used to inspire challenges to prejudice in daily lives. It is clear that the existence of ongoing safety issues means that Pride events, for some, mark a difference 'one day out of the whole year' where some LGBT people feel that they can 'announce who they are'.[8] Questions of visibility and outness are significant when understanding Pride events, which often perceive their historical roots as a celebration or re-enactment of the 1969 Stonewall riots in New York City. Although this trajectory has been questioned and the geographical specificity of Pride marches/parades needs critical interrogation, this imagined history creates collectivities and supports the myths of shared agendas, reconceptualising Pride events as sites of resistance to dominant (heterosexual) orders. Brighton Pride was for some an opportunity to create 'new worlds', promoting messages about the unacceptability of prejudice, abuse and discrimination.

The existence of ongoing prejudices was demonstrated in part by protests by Christian Right at the Brighton Pride parade. Yet, Robert argued even if 'they [the Christian Right] weren't physically shouting', the insidious and daily experiences of feeling different continued to be a factor in LGBT people's lives. Robert contended that Brighton Pride was a chance to 'show them they are wrong' through a 'positive celebration', making people feel ordinary for one day. Indeed the vast majority of LGB people questioned at Pride in Brighton and Hove in 2004 saw it as a celebration of LGBTQ identities (95 per cent n. 4399). Johnston (1998) noted the spatial importance of the parade marches, such that where the parade goes, gives the celebration political import. While much of gay Brighton was seen as safe, many areas that were 'passed through' by the parade were also perceived as aggressively and 'scarily' heteronormative (see Chapter 3). It is not just what is being celebrated, but also where this celebration occurs that questions and challenges heteronormative discriminations.

Challenging discrimination was built into the core remit of Pride in Brighton and Hove and this was reiterated in its charitable status. In 2004 Pride in Brighton and Hove became a charity led by a group of up to 12 trustees[9] with two to three workers (both part and full time). The organisation's charitable objectives were related to tackling prejudice and discrimination using the size and popularity of parade and park events:

---

8   This is not to negate the ways in which safety issues emerge during Pride events, where LGBT people can be targeted for abuse, including from those within the LGBT collective that it is supposed to be celebrating.

9   Kath was a Pride Trustee between 2005 and 2010. This positioning of course affects the writing of this chapter. Here we focus on the data collected, rather than her experiences.

> To promote equality and diversity and advance education to eliminate discrimination against the lesbian, gay, bisexual and transgender (LGBT) community.

> Pride raises awareness of issues by promoting and staging a series of events, including Winter Pride and the annual free summer festival, and making grants and/or donations to other charitable and voluntary organisations.

> Our key objective is to develop an environment in favour of LGBT equality by providing information, advice and support. (Brighton Pride 2011)

While Pride in Brighton and Hove aspired to the ideal of 'eliminating discrimination', this priority was not always actioned. Those who were involved in the organising of Brighton Pride in the latter part of the decade felt it was not doing enough to achieve its potential in this regard:

> Yvonne (ex-Chair of Pride in Brighton and Hove at the time of interview): I do believe that Pride needs to have a much stronger educational messaging, awareness raising, function, than it does at the moment. I think it has huge potential to do them a lot better and to achieve more, using the vehicle that this annual event gives it. I always come back to the point that Pride is a charity. It has very clearly stated charitable objectives. They are around education, preventing discrimination, promoting inclusion, awareness-raising about LGBT issues. I think that needs to go back to being the primary function and that some of the delivery mechanisms need to evolve to reflect that more. Its function as a celebration is also hugely important, but at the moment, my view is that its function as a celebration overshadows everything else. Don't get me wrong, it's absolutely essential that people feel able to come and have, even if it's just on one day a year, the most amazing time feeling totally included, feeling surrounded by people like themselves, feeling they can be themselves, but I think that should form part of, but not the primary, purpose of it.

Yvonne saw the potential of Brighton Pride to address some of the key questions that were pressing at the close of the first decade of the twenty-first century, but said that its 'function as a celebration overshadows everything else'. It is apparent throughout our data that critiques were happening from within the organisation of Pride in Brighton and Hove. As we have contended, the notion of what should happen in Brighton was influential to driving political action in the city, and Yvonne's narrative (as Chair of Pride 2007–2009) illustrates a desire to move from the party to more identifiable 'politics', recognising both the import of celebrating LGBT identities and the need to do more. However, as we shall see below, movement towards a model that was based less on partying and 'getting out of your heads en masse' was met with the imposition of gay male power from those who like it 'how it is' (and make money from this model). Pride events, as with all activist endeavours, are flawed and imperfect, but they need be explored for their positive possibilities, as well as critiqued for their normalising and commercial impetus. We will now further explore the ways in which Brighton Pride sought to include the diversity of LGBT identities, explicitly refusing singular unified messages that could have been exclusionary.

One Community, One Message?: Diversity and LGBT Politics

While the discourses of Pride losing its way could be in common with other Pride events in the Global North (and indeed locally focused discussions were informed by debates elsewhere), Brighton played a part in the politicisation of Pride in this city. This disrupts a straightforward narrative of protest marches that presume unity and common enemies, which are replaced with parades that celebrate, party and dance with diversities. In Brighton, oppositional politics were only one part of the canon of LGBT activisms and LGBT lives could not be read solely through a bleak picture of violence, abuse and discrimination. It is clear that the possibilities of celebration create an imperative to seek other ways of knowing, other narratives and other stories (see Sedgwick 2003).

Pride in Brighton and Hove did not have an overarching agenda in the period under discussion. While the parade was themed each year (Pride Around the World, Pride Beside the Seaside) and groups were encouraged to relate their float to the theme, there was no requirement to present a particular viewpoint or follow a specific line of thinking or protest. The absence of an overarching message arguably enabled diverse manifestations and expressions of LGBT lives and agendas. This approach sought to address the issues of creating one representation of acceptable gay lives and, in theory, all expressions of sexual and gender diversities were welcomed. This could be understood as a political move, because it worked to diversify the visible aspects of LGBT lives. In moving from singular agendas there was no one message set by more empowered groupings, such as powerful gay men or businesses and commercial interests. The lack of dictation about what to say or how to feel had the potential to diversify the projection of a homonormative version of LGBT lives.

Some of this potential was realised. For example, the Pride in Brighton and Hove parade in 2005 featured a float from Pride in Whitehawk (an area of social deprivation). which was decorated by LGBT and non-LGBT people, involving over 30 local community groups. After parading through the streets of Whitehawk, the float joined the Pride parade through the city centre.[10] Contesting the classing

---

10   This initiative was driven by a community centre in Whitehawk during the period of investment by national government in neighbourhood renewal areas. The group also produced a show as part of the 2006 Winter Pride festival in Whitehawk, featuring local LGBT 'celebrities'. In the 'golden age' of spending on social inclusion, a publicly funded community development worker in Whitehawk felt both able and supported to undertake

of scene spaces (see Chapter 4), it was understood that the initiative offered representation of LGBT people in these areas. It was understood to have educated not just non-LGBT people in Whitehawk, but also challenged stereotyping within the LGBT communities:

> Ed: The whole of Whitehawk came out of their houses and cheered and clapped [as the Pride in Whitehawk float went past]. They wouldn't have done before, a few years ago, if anybody had come out as openly gay. [Pride in Whitehawk] has definitely put Whitehawk on the map with [regards to] people's perception of it. There are a lot of closet gays here. I mean the buses are full of them. It's much better accepted by ordinary people, men, women and boys and girls of the street. Nobody hurled stones or abuse, it was a happy day.

> Michael: the Pride groups go down the [name of the central road in the area] on a great big float on a Saturday morning going to Pride, just challenging people's idea of what being LGBT is. Same as what's challenging some idea, if you go to a pub and you say I'm from Whitehawk and they're like 'well where's your Burberry?', just challenging that same thing. So I think it's just about having more positive imagery, about making sure that you're doing things like we're doing so that more people can see. It's challenging what they usually think about LGBT community. (Pride in Whitehawk focus group)

Because there was no direction about unified messages, people who got involved in Brighton Pride could offer their own. Explicitly or incidentally these creations in turn created the diversity of the parade, festival week and park event. This was underpinned by an ideal that Pride in Brighton and Hove could create a platform, where diversity was represented. In this way, an archetypal gay was only one image representing LGBT lives. The plethora of representations and autonomy of messages in Pride in Brighton and Hove parades meant that these were not uniform or predictable even over the course of one event.

Through Pride, Brighton was represented as a diverse place where LGBT people were not homogeneous, and this was important when defining the charity's agendas. Such a model contrasted with other Pride events where individuals were appointed to represent specific 'communities', or those who organised the event set out predefined agendas based on assumptions about what LGBT people care about and around which they should unite:

---

this work. With the loss of funding for areas of social deprivation and the subsequent loss of this worker, the Pride in Whitehawk group did not continue. It could not survive with only volunteer input, regardless of the interest and excitement that surrounded the group, demonstrating that state funding can be crucial for developing LGBT initiatives that were not commercially-oriented or viable.

Robert: Pride [in Brighton and Hove] doesn't have a tangible objective that it is there to promote. It is there to give a platform to the smaller groups who don't necessarily have a platform or a voice to say what they want to say. I would never want Pride to speak on behalf of those groups. That's always difficult – people want to know what Pride stands for. We have our fundamental objectives of fighting the phobias and celebrating LGBT lives, but we do that through providing a platform for these other smaller groups to say what they want to say. It's very difficult to try and explain that to some people cos they just want us to give a political statement and we can't cos it's the [name of trans project] statement or [name of LGBT youth group] statement, it's not ours, and I wouldn't deem to talk for them.

Local government support for the event came not only in the form of tourist development initiatives, but also through financial support facilitating the involvement of grassroots LGBT groups. For example, in the first decade of the twenty-first century Brighton and Hove City Council offered grants to groups to pay for the costs of decorating floats, providing costumes and other expenses and provided funding for tents and stalls on the park. This sought to help diverse groups, including marginalised LGBT people, to be part of the parade and present in the park. Pride in Brighton and Hove (the organisation) made floats available to local LGBT groups for reduced rates and had a scale that put the majority of costs onto non-local commercial enterprises. They sought to ensure affordable parade entry to these groups (which included workshops designed to facilitate creation of walking tableaus and floats) and offered grants to groups to put on events during the Pride week. It contested the drive to invoke a message, which would be selected by a few, and instead worked to accommodate diversity, difference and multiplicity.

Although Brighton Pride's organisers sought to make the event a vehicle for LGBT people and groups to present their messages, how such messages were read was not controllable (see Johnston 2006). At times these readings reflected some of the internal prejudices between L, G, B and T (see Chapter 5) and questioned the collectivity of LGBT people who were supposedly celebrated through these events:

Eve: There were quite a few of us [from Brighton Bothways, who went on the parade] the first time. I was quite nervous about it. We wondered how we'd be received by people. [The reactions were] good and bad. We had people cheering like they were really proud of people that were actually out, within the parade, we had really positive support. Then we had other people, going 'fucking greedy bis,' and all that kind of stuff, along the parade. It was like, 'Really? On Pride? On one day, you can't be supportive?'. So in some ways, it was really great and in other ways, it was really depressing.

Eve's narrative points to the contested ways in which LGBT identities can be received in a vehicle such as Brighton Pride. Both reactions have political relevance, either through challenging hegemonic understandings of who is welcome at Pride and who is a part of the LGBT collective, and also demonstrating solidarities with those who are often marginalised within the category LGBT. The politics of Pride parades are reconstituted in part by both those who created the floats and those who viewed and responded to them.

Diversity has been a key part of, and tension within, Brighton Pride, even when there was supposedly community cohesion and distinct lesbian and gay causes in the golden age of protest politics of Pride. Where cohesion and overarching agendas existed, this created exclusions and the valorisation of specific people and messages. Although by the middle of the first decade Pride in Brighton and Hove actively sought to include marginalised LGBT groupings, during the early part of the decade this was a hard fought battle with those running Pride:[11]

> Dana: There was a group of BME people that started the BME Pride Tent, which was brilliant. I don't remember clearly and I don't want to be unfair, but I don't have the sense that it was easy to get the tent. That was an emergent process as well so Pride organisers asked 'What exactly are they doing?', 'When exactly are they going to do it?', 'How exactly are they going to get the money to do it?'. We weren't particularly organised so we didn't get things in on time for the tent. But the tent wasn't welcomed; it was grudgingly allowed and nearly not allowed. That's what it felt like and again it was like 'God, what are you going to have next? A disabled black lesbian's tent?'.

The presumed gay majority clashed with the inclusion of raced classed 'others' who sought specific space not only on the parade, but also in a park event. In providing a platform, Brighton Pride organisers framed how others could participate in Pride events. Dana (who was Spectrum coordinator when facilitating this BME tent) shows how the cultures of Pride clashed with the emergent process of putting together a BME Pride tent as a new venture.[12] Pride cultures developed in response to the necessity for deadlines, organisation and professionalism (see Newman and Clarke

---

11   Spectrum, see Chapter 6, not only contested the whiteness of the festival seeking a space for BME people, later in the decade it challenged Pride in Brighton and Hove for its lack of inclusiveness with regards to disability access, creating space in the park (an access tent) from the mid part of the decade that was valued by many of our participants.

12   It was not only Pride in Brighton and Hove that needed to be convinced of the values of separatist and dedicated spaces, straight feminist politics in the city also made creating women's space in the park controversial. Sheila also spoke of the controversy with the Women's Centre who were a 'little bit baffled' and her story demonstrated the 'hard slog' of getting a women's tent as part of the park. This had changed by the time of the interview, where the Women's Centre had become much more supportive and recognised the importance of the women's tent.

2009, Ward 2003) in order to put on an event with [at that time] 70,000 people. Groups were required to work within parameters of the park and parade necessary for coordinating a large event safely. This inevitably privileges some LGBT people over others, casting some as part of the event and others as outside of it. Thus, the ideal platform for LGBT people to represent themselves was located within structures that were focused on the practical issues associated with the delivering of events. This was problematic where individuals and groups were feeling stretched, or marginalised LGBT people lacked networks and competencies to undertake this work.[13] Undoubtedly there were groups and individuals who did not participate because of the structures imposed by this model. As Lee acknowledged, the ideal of self-organising your involvement in Pride was important, but the realities of such organisation in terms of capacity, skills, knowledge, confidence and feelings of belonging were for some prohibitive. Taking up these opportunities to be part of Pride required particular forms of cultural identification and resources, which not all LGBT people or groups in Brighton possessed. In operating a model where Pride in Brighton and Hove was a vehicle for messaging and, by offering a platform of park, a parade and a week-long series of events, the 'organise it yourself' ethos could, ironically perhaps, be disempowering for those who most needed a platform for their messages to be heard.

Imaginings of Brighton as a large gay metropolis and Pride in Brighton and Hove as a core aspect of this meant that there were expectations of capacity within the organisation. However, the realities of a small city failed to fully attain the hopes for inclusion promised by Pride as a platform. The organisation of Pride of Brighton and Hove (trustees and workers) was perceived to have additional capacity that 'should' have allowed it to gather resources for specific tents (this was particularly mentioned by organisers of the women's tent). This myth was dispelled by those who volunteered for Pride:

> Yvonne: There is a public perception of an organisation that is significantly more well-resourced and funded and skilled than it actually is. I think it has challenged me and others, because the biggest issue to move and change organisations [is] you have to have the resources to do so and you have to be able to take people with you. Pride suffers in both of those respects because it is representing such a diverse community, within which there are numerous different views and opinions, and some of whom shout an awful lot louder than others.

Throughout the first decade of the twenty-first century, Pride was run by five to ten committed volunteers with the support of (at its peak) three members of staff. As chair of Pride, Yvonne worked to change the focus of Brighton Pride (to make it more community and educationally orientated, in line with the charitable

---

13   From 2006–2009, workshops were run on 'how to participate in Pride', drawing on the expertise of those who had done this in the past to build community capacity, with training courses on other aspects (such as press coverage).

objectives of the organisation). She also gave attention to maintaining viability by working within the resource limitations of the organisation. However such long-term strategic thinking was often side-lined in favour of getting the event done, with a focus solely on 'this year'. The inclusions of the platform of Pride and the exclusions of the structuring of a Pride event were inflected by the place of Brighton, where it was expected that a large event would be well resourced and financed, given the assumption of Brighton's 'large LGBT community'. We now move to explore in depth the contestations around the commercialisation of Pride in Brighton and Hove, which offers an alternative perspective to this supposed capacity.

Keep Pride Free or Selling Out?: Commercialisation and Politics

In the early part of the decade, Pride in London moved to charging an entry fee for its after-parade event. The event was roundly critiqued and boycotted for this move (see Brown 2007). Unlike Manchester, which has charged for entry into its event since the 1990s, such commercialisation was seen to question the very tenets of London Pride, and was seen as an apolitical move. A key issue in the use of consumption to advance lesbian and gay politics is the exclusion of those who cannot afford to buy in, as well as the racialisation of particular forms of consumption practices (see Markwell 2002, Puar 2002, Kates and Belk 2001). Here we seek to consider the nuances of 'commercialisation' and the possibilities of 'commercial' politics. Once again, this is not to deny the importance of contesting the deployment of repressive neoliberal practices, nor to negate the critiques of commercialisation. Rather, we want to highlight the messiness of equalities landscapes that attend to these critiques, and also recognise that there is an excess to them. Our key contention is that while Pride in Brighton and Hove was not anti-capitalist, this does not *necessarily* make it apolitical, when we allow for a plethora of political possibilities.

There is no unified understanding of what makes a Pride event 'commercial', indeed in the CMIT research we identified at least three understandings of commercialisation: paying to attend; making money for businesses; and involvement of business-focused people. Firstly, up to 2010, Pride in Brighton and Hove offered a free event in one of the city's parks. Pride in Brighton and Hove used the phrase 'the biggest free LGBT festival in Europe' and this was seen as unique, inviting 'anyone to come and celebrate those LGBT lifestyles' (Robert). The term commercial was often equated with paying to attend the festival and, as such, Pride in Brighton and Hove was understood as not commercial. However, and secondly, this contention was challenged by the view that Pride in Brighton and Hove was 'commercial' because commercial enterprises were involved and contractors, venues and other businesses were making money. This was partly on the back of the free labour that was contributed by those working in a voluntary capacity to ensure that the festival occurred. This offered a different perspective on the commercialisation of Pride as 'businesses mak[ing] loads of money' (Sheila) and, for some, this constituted 'selling out'. Finally, commercial could also be associated with individuals who were involved in the organising committee. While the interests of those running Pride altered over the decade, memories of 'commercially minded people' and the resulting 'loss of confidence in the concept' (Nigella) remained, and defined Pride. This understanding linked individual personalities and perceptions of Pride as 'too commercial'. Such diverse conceptualisations of the commercialisations of Pride are important to highlight, as commercialisation is often used as a homogeneous and supposedly unified trope against most large scale Pride events in the Global North. Yet even between our key informants there was little agreement about what commercialisation might mean and whether it might pertain to Brighton Pride.

Kates and Belk (2001) point to how historically gay rights have been framed within Marxist ideologies. Thus for some, commercial ventures, such as sponsorship, result in a loss of key meanings of Pride events and appear irreconcilable with these events or strip them of their subversive roots. Even where Pride events do not charge, the perceived loss of 'leftist ideas' can be seen by some as a commercialisation of the festival. As we have seen, Brighton Pride did not have a unified, identifiable enemy to protest against, and the organisation actively courted commercial sponsors to enable it to run. Therefore, for some, the event itself became something that needed to be opposed, as commercial, assimilationist and colluding with state powers. With commercialisation understood as more than paying for the park event, actions that protested against Pride[14] challenged the commercialisation of Brighton Pride, in multiple forms:

Annie: [I am] someone who has protested against Pride at Pride. I joined [Brighton] Queer Mutiny almost precisely so that I could help work to re-politicise [Pride]. It was the year that the army marched at Pride and it was during when we were still in Iraq. We all suited and booted up and kind of wiped the floor behind the army with mops and [were] quite roughly treated by the stewards and stuff. In following years Queer Mutiny had brought political banners to Pride. They've always been asked to take them out. I don't think Pride's very sympathetic to the idea that there's a politics of LGBT in Brighton and Hove. What it seems to me to want to do is to celebrate LGBT without really

---

14   One example of this is 'gay shame' in London which defines itself as 'a creative rebellion against the banalities of the mainstream [London] Gay Pride festival and a satire on the commercialisation of our community' (www.duckie.co.uk, 2010). In contrast, Pride in Brighton and Hove in 2009 supported Club Wotever Brighton, a local Queer group that sought to create DIY events that queer gender and sexuality, illustrating the importance of taking place seriously when exploring political action. 'Prides' are not homogeneous, even with the entwined networks of London and Brighton. This also demonstrates the possibilities of Pride as a platform that was created by a range of diverse LGBT people, rather than 'something' that organises events 'for' people.   .

ever questioning what that means and who's involved in it and what it means for the people who are involved in it, and to get buy-in from corporate sponsors and from institutions in Brighton and Hove. It's you know, 'Pride sponsored by insert your sparkling wine company here'. Seriously.

In 2007 Queer Mutiny demonstrated at Brighton Pride, protesting both about Pride's sponsorship deals with Barclays ('Buggery not Barclays') and the inclusion of the police on the parade ('Cunts not Cops') (see Figures 8.1 and 8.2, see also Brown 2007).[15] This sought to re-politicise Brighton Pride by challenging the selling out seen to be manifest through corporate sponsorship and collusion with 'repressive state apparatus' (in this case, the army). For Annie, LGBT politics cannot be aligned with celebrating LGBT without reflexively considering the effects of sponsorship and the inclusion of the state. More than this, she asserts that Brighton Pride had a core message and that it was lost to sponsorship ('insert your sparkling wine company here'). The Queer Mutiny protest banners depicted in Figures 8.1 and 8.2 were removed from the parade half way along the route by the police. Police claimed their banners were offensive, emphasising perhaps a protest and parade divide and the power of the police in regulating mass events such as Pride (as well as ironically reiterating the message that Queer Mutiny sought to get across regarding the oppression of protestors by the police). This also points to collusion with the police and commercial sponsors by the Pride organisation.[16]

However, the binary of collusion and resistance can assume a particular mutual exclusivity between consumerism and politics based on the opposition between these. Waitt and Gorman Murray (2008) note that sexual politics can be reconfigured through markets.[17] For some, politics cannot engage in commercialism or capitalist ventures, however the linking of sexual and gender liberation to anti-capitalism also needs critical interrogation. Perhaps politics can take diverse and complex forms, some of which might not be seen as 'political' by those who understand LGBT/queer politics as always and necessarily outside of corporate buy in.

---

15   Ironically, Christian Voice, a right wing far-right Christian group, were also demonstrating on the parade route and also in protest to the police heading the parade.

16   Although a coherent decision such as this was not possible in the milieu of the parade event. Illustrating the disparity between 'the organisation' and what happens at 'the event' – at subsequent trustee meetings, some trustees supported the police/security decision citing the lack of 'family friendliness', others pointed to the need for Pride to enable all who wanted to have their message on the parade to be enabled to do this.

17   They point to Daylesford's 'Chill Out' festival to contend that the festival disrupts as well as reinforces straight/gay binaries. Pride in Brighton and Hove in and of itself has to offer both community engagement and 'political' meaning in order to garner and retain financial support through (gay) business involvement as well as other sponsors 'corporate social responsibility' funding. The segregation of the political and the commercial is thus not possible in this context.

Figure 8.1 'Cunts not Cops'

Figure 8.2 'Buggery not Barclays'

Charges of commercialisation often assume specific forms of corporate sponsorship, and Pride in Brighton and Hove appealed specifically to 'corporate social responsibility' funds. There is little work that examines the motives for sponsorship of such festivals and the effects that such sponsorship has on organisations, employees and others associated with it. An examination of corporate cultures might find that businesses have nuanced motivations for the support of LGBT events, with potentials for unintended challenges to hetero- (and indeed homo-) normative orders. In signing up to Brighton Pride, they became part of a festival that celebrates LGBT lives and this could be read as a political move; it could, for example, challenge the heteronormative cultures of an organisation. Studies that focus on consumption and commercialisations of Pride festivals reaffirm the importance of the political symbolism of the festival over the economic factors (Brickell 2001, Kates and Belk 2001, Waitt and Gorman and Murray 2008). It is clear that Pride in Brighton and Hove's politics were not anti-capitalist, they colluded with the state and business interests, however this does not mean they were only or *necessarily* apolitical. In particular, taking place as a constitutive factor, Brighton Pride's celebration of the diversity of LGBT lives and platform for diverse LGBT representation offers potentials for activisms, only some of which were realised. Thus, there is an excess to reading Prides as apolitical because of association with corporate sponsors and the state. Instead each Pride event needs to be taken on its own territorial terms, while also recognising the overarching critiques that hold salience.

The politics of Pride in Brighton and Hove could not manifest the ideals that were desired by LGBT people, such as affecting sexual and gender liberation for all. Brighton Pride did not necessarily produce institutional change even amongst its supporters. Cynicism about the presence of the police and army in the parade also extended to the council's involvement in Brighton Pride (see Frank above), associating it with tourist revenue, but not catering for LGBT people:

> Steve: The gay community probably draws more tourists here than any other section and even the pier. If we had both piers up at full swing I don't think anything could beat Pride Week for the money that we produce for Brighton Council, and they treat us like dirt. (Hate Crimes Focus Group)

Pride was used to court LGBT communities by politicians, as well as being seen as an important economic driver in the city. Yet, some LGBT people felt that Brighton Pride events did not necessarily foster social change in local government practices. In this dissonance between experience and what Brighton Council *should* be doing, Steve was able to make a claim for better treatment on the basis of solidarity with 'a community that draws in more tourists than any other section'. Brighton and Hove City Council was diffuse and diverse, where one department (tourism, or indeed individuals within it) might be open and accepting, while another, in this instance housing and safety services (see Chapters 3 and 7), might be disempowering and oppressive. Consequently, Brighton Pride provided a basis for making a claim, but the claim was not necessarily upheld through council support of Pride festivals. Brighton was not only created through Pride, the city of Brighton created rights claims through hosting and supporting Pride in Brighton and Hove and being closely aligned with its 'success'. There can be little doubt that place relations were diversely manifest and experienced, such that the benefits of Brighton Pride affected LGBT people differently across the collective.

Ordinary activisms were manifest through Pride in Brighton and Hove despite, or indeed because of, commercial involvement in the festival. Yet, as we have seen throughout this section, the promise of ordinary activisms were only partially achieved and hegemonic relationships can be recuperated. We now move to question the conflation of lesbians and gay men, and discuss how gendered power relations were recreated.

Gendered Power and Gay Politics

Pride events are often spoken about in terms of the events themselves, the parade and after-party, yet interrogating Pride organisations can also offer insights into the nuanced operationalisations of power within LGBT communities. As these events acquire economic, political and social significance, in addition to the widening identities of who is part of Pride, contestations regarding the ownership of Pride may be even more fraught. In the context of gay Brighton, examining the 'power of Pride' offered insights into the gendered hierarchies that created and controlled aspects of gay Brighton.

Since its beginnings in 1992, the organisation of Brighton Pride has been fraught and contested. These contestations reproduced power relations identifiable within LGBT collectives, most notably between men and women. The desire to keep control of Pride was a key facet of maintaining (and creating) gay power in the city. Thus, power was not transferred uniformly to LGBT people who are nominally given responsibility and indeed accountability for Pride in Brighton and Hove. Rather, gay men played a significant role in supporting and tearing down those they perceived to be not doing a good enough job:

> Paul: The two mad lesbians that ran it in the early 2000s. Everybody was going away from Pride and these two women were still in the middle. [Woman's name 1] hadn't the organisational skills to pull this event off and she would just threaten people. We had the big public meeting. She was called to the meeting and everybody turned up. We had 80 people in there. It was a dreadful meeting because they were all after [woman's name 1]. She was fighting for her life and she just didn't want to let go.
>
> The reclaiming of Pride, in 2002, what happened [then], [woman's name 2] and [man's name 1] – I ran the campaign to get rid of them. My lowest point has been three conversations, all to do with Pride. When I told [woman's name 1] she had

to go, when I told [woman's name 2] she had to go and [man's name]. I had run his campaign to get him elected as chair of Pride because I thought at the time he was the right person for the job. Then I said to him, 'you've done enough for a year why not pass it on to somebody else' and he said, 'oh well it can't happen without me'. I walked away from a situation I would normally have dealt with.

[Woman's name 2] was Chair of Pride and then at the AGM she announced that she was going to work for the [name of business] and she got a very bad reaction. Everything was going downhill, nobody wanted to work with her. I said, 'I really think you need to go, darling, because nobody wants to work with you and if we have another fight like we did with [female name 1] it's just gonna be very damaging for the organisation.' She stepped down.

I've had no problem with going along with being the bad guy and then I'll walk away from it and not worry about it. I don't think I've ever done anything to harm anybody gay wise in Brighton and I've always thought that I've fought a good case for fighting for the underdog.

We quote Paul at length to illustrate the complex histories of politics of Brighton Pride through his eyes. This passage demonstrates his power and influence, controlling what he saw as the right direction for Brighton Pride (in part through discourses of 'everybody' and 'nobody'). Indeed, when his agenda was not pursued, media outlets and powerful gay men engaged in campaigns to oust people, both publicly and in private 'backrooms'. In this way, control of Pride was exercised through gay male networks. The invocation of community discontent to 'get rid' of two women from Pride is illustrative of the ways in which these campaigns were pursued. While [male name 1] was also a contentious figure and deemed by Paul to be no longer fit for the job, he was not dispossessed by Paul who 'walked away' rather than using the tactics he describes when dealing with the women who occupied similar positions. During [male name 1]'s time at Pride, the organisation became a charity, with new structures that removed the possibility of people coming to meetings with supporters and getting elected as chair of Pride. The size and scope of the festival increased and the organisation gained enough funding to employ staff. Structures were put in place that made it more difficult for individuals to gain control of the organisation. These changes reduced the extent to which other gay men, including Paul, could control the organisation of Brighton Pride.

Women spoke very differently about the removal of women from powerful positions, including in the organisations that ran Brighton Pride:

Louise: [A] low time was when Pride was being chastised for the work it had done. It just felt like a direct attack on women again. There were obviously some issues that needed to be brought to light, but that was really awful. Just really soul destroying for a lot of women in the city. It felt like it was an attack because she [woman's name 1] was a woman, rather than some of the factual issues around Pride. There seems to be a bit of a trend that goes on in the city. [It] seems to go in troughs and peaks. So you'll get women doing some really amazing bit of work and, because it doesn't work in the way it's wanted to be worked, it tends to get collapsed by small numbers of quite powerful men. In Pride twice I think it's happened, some really amazing women who got really, really hurt in all of that. The current situation [2009] that Pride finds itself in, again there's quite a high number of women involved, again it's that sustained attack.

In contrast to Paul's perception ('I don't think I've ever done anything to harm anybody gay wise in Brighton'), Louise purported that women who gain powerful positions could find that their work was 'collapsed by quite powerful men' and that the women get 'really, really hurt'. Such internal conflicts have detrimental effects on people's lives and Louise's narrative suggests that it is women who experience this disproportionally. Thus, along with organising a key event and being part of an important organisation which had the symbolic importance of being 'in control' of Pride, there were detrimental social effects.

Where women were subject to sustained attack and removal from key positions, the gendering of gay Brighton is apparent. Pride, as an organisation, was also deployed by powerful men in ways that were detrimental to LGBT women:

Dana: I came up against the gay male power contingent which was quite bruising. It was Pride actually. I felt as though it was dominated by powerful men with lots of money who were misogynistic. I found it quite frightening actually. But [male name 1] came in and was very charming and we had a good long chat. I was talking about Spectrum and at that point I was feeling quite eroded and quite unsure of myself and I talked to him really honestly about it. I was saying 'Look I'm not sure what Spectrum should be doing and I'm not sure what I should be doing' and he seemed really sympathetic and understanding. But apparently he went straight to Paul and said something along the lines of 'She's rubbish, she's not going to last, Spectrum's hopeless'. I didn't know that for ages. I found out a year and a half later. Paul told me and I felt really, really hurt.

Pride asked to move in to the Spectrum offices and it felt like they took over. The space became noisy; there were masses of people and masses of things happening, the room was taken over. So I then retreated to a tiny room at the back and it felt like it signified a huge amount because there was this massive room which was very tranquil and very respectful and very peaceful and then suddenly it was filled with money and men and business. It felt like an empire building force just came in.

So I thought it could be good about [name of Dana's successor from 2005] coming. I thought, 'Well maybe it does need a gay man, actually, Spectrum, maybe that's the only way it's going to work.'

Dana paints a vivid picture of gay male power being deployed through an 'empire building force' that invaded space and brought 'masses' of people. This mirrors Paul's language above where he refers to a 'big public meeting' where 'everyone turned up'. Paul told Dana of [man's name 1]'s betrayal of information that she had shared in confidence at the same time as Paul was trying to remove [man's name 1] from his position at Pride. The 'bruising' nature of these encounters was apparent throughout our data, and particularly from women, illustrating the hurt endured when participating in LGBT community work in gay Brighton.

In contrast to the 'empire building' discussed by Paul and Dana, Robert and Yvonne held a different understanding of the role of Pride trustees and Chairs. This was rooted in the notion of Pride as a charity, with charitable aims and 'professionalised ways of working'. Charitable status limited the gains that can be accrued by trustees. Yet Pride in Brighton and Hove, as a charity, came under sustained critique by facets of the local gay media because charitable status means that Pride was accountable to the charity commission and was not portrayed as relating to the 'local gay community'. However, the deployment of 'gay community' was often linked to gay male control of gay Brighton, in ways that did not encompass community in the ways implied by the use of the term.

In 2010 the power-play of Pride came into sharp relief and contestations regarding community were manifest. Pride in Brighton and Hove declared a deficit and outlined plans to scale back the festival, stating these would make it more 'community orientated', more focused on its charitable objectives and sustainable as a free event. These proposals would have reduced the number of (commercially based) tents in the park event that were significantly supported by the core funds of the organisation. There was a significant backlash through aggressive confrontations at meetings and social media. The resistance to change came from those who said they enjoyed Pride 'as it is' and did not want what they termed a 'picnic in the park',[18] as well as from businesses whose large tents had been heavily subsidised by sponsorship brought in by Pride in Brighton and Hove.[19] The resistance to these changes, similar to 2002 (see Paul above) was entitled 'Reclaim Brighton Pride'. After several consecutive years of being awarded the contract to service the park, Pride in Brighton and Hove put the contract out for competitive tender. The leadership of the disgruntled production company then drew up a 'Rescue plan' that deployed discourses of 'gay (and

---

18   The idea of keeping Brighton Pride 'the same' (unchanged) was conflated with 'Save Brighton Pride' (name of Facebook site), deploying a range of different perspectives described above simultaneously and perhaps contradictorily.

19   There was also a significant resistance to the tendering processes that placed longstanding contractors in jeopardy if they could not provide the services required or (as was the case of the main park contractor, who subsequently also lost control of London Pride) they did not bid for the contract.

more rarely LGBT) community'.[20] It was clear throughout the debate that Pride in Brighton and Hove, a powerful organisation in the city, was not playing by the implicit rules of an invoked 'gay community'. A focus on a more community-oriented event undermined the power of the 'homonormative' majority, who felt they held 'ownership' of Pride, on behalf of the 'gay community'. This led to a heated meeting, where it was perceived that 'a lot of gay business men [were] shouting at a number of lesbians who were trying to make Pride happen' (Dana).

In the period from 2000, this is the first occurrence that we can identify where the control of Pride did not go back under the control of those pushed by the local gay media with interventions from particular gay men. In part this was because the charitable status of the organisation prevented 'vested interests' from gaining decision-making powers. 'The gay community' was mobilised through specific tropes and certain groups, whose activities focused on Pride events stood in for 'the gay community'. In order to 'reclaim Pride' and gain control, gay men (in the main) who had access to resources such as the local gay media deployed oppositional politics that spoke of acting 'for' and 'against' 'the gay community'. They also spoke of themselves and their allies as 'everyone' and as 'representative of the gay community', with more legitimate claims to Pride than other LGBT people.

The attempt to reclaim control had familiar effects on those who volunteered (as well as workers), and once again women were in the frontline:[21]

> Yvonne: My low points have also absolutely [been] being involved in Pride and the developments that took place which resulted in my resignation as chair. I think they are an appalling statement of a very small section of the LGBT community, who have a particular way of doing things, which makes it very difficult for other people sometimes to bring about change, because a small minority will always block it. It's about people whose behaviour is seriously shameful in terms of the way they personally attack other individuals within the LGBT community, who are trying to actually do good work, because it doesn't suit them. What I saw happening to people around me who were involved in doing good things for the right reasons, without self-interest, but who were attacked personally. There probably will not be ever a strong enough voice to stop that from happening, and it does happen every few years unfortunately, when particular interests are being threatened.

---

20   This bid talked about 'Gay Pride' as a cutting edge intervention reclaiming the term 'gay' without acknowledging its exclusions. It restricted entry to those who could pay and asked for donations on top, and was a really poor deal for teenagers: under 12s admitted free with an adult and 12–18 year olds paying but with no entry to dance tents. Throughout youth culture was equated with homophobia (*Gscene*, 2011).

21   Indeed when the female Chair of Pride resigned, because of the abuse she was experiencing, attention did not deflect to the new male Chair. Instead 'Pride' became faceless, until the one full time female member of staff was verbally attacked.

Yvonne's narrative rings true with other women who have been involved not only in Pride but other LGBT ventures in the city on a voluntary basis. The return for many women was not 'kick backs', rather metaphorical 'kickings'. Thus, although some gay men and lesbians may have become normative and it is common to put this category together in discussions of homonormativity, focusing on Pride illustrates the importance of retaining an appreciation of gender differences when examining normalisations and inclusions. There can be little doubt that Brighton also played a role in reiterating gay male power. As a city where some LGBT people feel sorted, a small minority can capture and hold control, representing themselves as 'the gay community' who has to be recognised and catered for by the city. Of course these forms of control are always precarious and, as we have seen, there was a periodic reassertion of gay male control of key organisations such as Pride in Brighton and Hove.

Conclusion: Rethinking What Is and What Can Be Political

Our conceptualisation of ordinary activisms asks for a reconsideration of the ways in which activisms might strive for ordinariness. In this chapter we explore a festival which initially sought to highlight differences between them and us, and now seeks to celebrate LGBT people. The ordinariness we addressed then pertains to the ways in which LGBT people feel ordinary for one day, while continuing to recognise and celebrate differences. Taking place as a constitutive factor in the politics of Pride, in Brighton organisers worked towards including marginalised LGBT people and rejecting normative overarching (and exclusionary) agendas, creating spaces where LGBT people might feel ordinary, even if they are represented as extraordinary.

Of course the ideals of including diverse and marginalised LGBT people alongside those who seemingly represent homonormative ideals were only partially achieved. As we have seen with other forms of ordinary activisms, Pride in Brighton and Hove illustrates the complexities of debates regarding LGBT politics. We sought to add nuance to the charges of 'apoliticisation' through the 'commercialisations' of Pride events. We have argued that Pride in Brighton and Hove's politics were not anti-capitalist, but neither were they necessarily apolitical. However, as we have also illustrated, the hopes for making a diversity of LGBT people ordinary, even for one day, are only partially attained.

As we have seen throughout this book, place was diversely experienced. Focusing on gay male control of this central event, we have considered how the intra-LGBT politics need to be accounted for in discussions of homonormativities. Too often lesbians and gay men are conflated negating the gendering of homonormativities. Exploring the vilification of women who stood in positions of power, such as the chair of Pride in Brighton and Hove, illustrates the ways in which gender continues to matter.

Thus, it is not the case that all Pride events are necessarily political or apolitical, rather each event and organisation needs to be examined accounting for the constitutive relations of place. In the concluding chapter we begin discussions of the implications of ordinary lives and ordinary activisms, developing understandings of commonplace.

Chapter 9

Ordinary in Brighton?: Conclusion

[W]hen I think of why I love Brighton, it's The World's Least Convincing Transvestite that springs to mind. A man who has made the bold fashion decision to sport a jaw-dropping combination of earrings, eyeshadow, stubble and shaving rash on a daily basis, he has all the bewitching femininity of a rugby league prop forward in a pencil skirt. Judging by his clothes – demure court shoes, tights, pussy-bow blouse – he's en route to a clerical job in an office. For all I know, he might be facing yet another day of bruising homophobia and derision from his colleagues, but it doesn't look like it. He just looks like an ordinary bloke on his way to an ordinary job, albeit dressed as a woman. I've got a sneaking suspicion his workmates just let him get on with it. And if they do, that would be very Brighton. (Alexis Petridis, *The Guardian*, Wednesday 19 May 2010)

Introduction

Examining urban life for lesbian, gay, bisexual and trans people has enabled us to explore how equalities legislation was enacted through the possibilities and limitations of inclusions. Geographers have long conceptualised the intersections between sexual and gender identities and place. In this book we have shown how LGBT lives and activisms are informed by city imaginings, in this case, the 'liberal', 'tolerant' city that forms the basis of the idealisations of Brighton. This is evidenced when Petridis above points to a visible 'transvestite' as part of the supposed ordinariness of LGBT people in Brighton. Yet the person he discusses is still worthy of comment, and he uses male pronouns and derides this person's femininity through a supposedly incongruous use of male and female signifiers. He exemplifies the idealisations of Brighton and their failures.

Petridis's piece relies on aspects of Brighton that we have explored in this book. We have explored the presumption of tolerance and acceptance for all LGBT people in Brighton and looked at how this contrasts with lived experiences of the city. Labelling this person as 'an *ordinary* bloke on his [*sic*] way to an ordinary job, albeit dressed as a woman' with workmates that 'just let him [*sic*] get on with it', Petridis points to the imaginings of extra-ordinariness of the city. LGBT lives *should* be ordinary in Brighton and, in this way, *should* be better here than in other places. Here, LGBT people were believed to be part of the fabric of the city. Consequently, the exceptionalism of the city and the supposed ordinariness of LGBT people were mutually constituted. This ideal, while often reiterated, was for some LGBT people only partially attained. Furthermore, key forms of LGBT activism in the city were underpinned with the notion that Brighton *should* have been leading sexual and gendered inclusion agendas. In part because of the failure to achieve the idealisations of the gay capital, LGBT activists were able to work with public sector bodies in ways that sought to make LGBT people part of these institutions. To address activisms that strived to make LGBT people's lives ordinary, it was imperative to think about the city as a whole, including but not limited to gay areas (see Brown 2008).

Discussions of ordinary cultures, lifestyles and people have been an important pursuit of Cultural Studies, contesting a focus on 'high' culture, art and media through examining the routine aspects of everyday lives (see for example the work of Berlant 2011, Williams 1958, Bell and Hollows 2005, Highmore 2002, 2010). Here we are interested in the possibilities of pursuing ordinariness as a political endeavour. This chapter begins by highlighting the importance of critiques of normativities, before attending to the possibilities of ordinariness. We then argue for a politics of ordinariness, which seeks to create commonplace lives in ways that are not necessarily normalising. We finish by reflecting on the potentials of research following changes that were anticipated, and feared, at the conclusion of Count Me In Too (2010).

Normativities and Ongoing Marginalisations

At the start of the book we outlined how discussions of normalisation have critiqued LGBT equalities and the 'gains' of legislative change. Where equalities legislation is associated with adhering to normativities, these 'gains' are conforming and limiting. Discussions of homonormativities have offered important and fruitful insights into the limitations of such inclusions including the idealisations of particular relationship forms, the disciplinary power of hetero/homonormativities, the racial or classed production of sexual identities and how nation states justify the use of violence through discourses of liberation and progress (see for example Erel et al. 2011, Kuntsman and Mikaye 2008, Puar 2007, Taylor 2009, Warner 2002). In this vein, scholars have critiqued state withdrawal from public service provision and the commercialisations and commodification of gay lives (Duggan 2002, O'Brien 2008, Sears 2005). Pointing to other possibilities that might challenge these normalisations, queer theorists have explored the exceptional, abnormal and resistant, such as activisms that challenge commercial festivals like London Pride (see, for example, Brown 2007).

However, these discussions have limitations, including being lost in the ubquitousness of critique, leading to hopelessness (Sedgwick 2003), creating what Halberstam (2011: 1) terms 'cynical resignation' and negating the geographies that are inherent to the construction of discussions of homonormativities (Brown 2012).

In the CMIT research, some LGBT people experienced that being part of and cohering to 'LGBT' brought them within certain state recognitions and equalities initiatives. Some did become ordinary, with their lives in the 'gay capital' being part of the fabric of the city. They experienced acceptance in workplaces, enjoyed socialising both on and off the scene and felt that any needs that they might have would be met within the city.

Yet, while some people's experiences were characterised by unremarkable-ness, inclusion and acceptance, others talked of painful experiences of exclusion. Not all who come under the category lesbian, gay, bisexual and trans can access the subject positioning of ordinary. While some may not want to, others held expectations of inclusion for all LGBT people. These were juxtaposed with continuing experiences of exclusion. We found that many of the 'old' problems of exclusion, marginalisation and abuse continued to exist, manifesting in poverty, housing and welfare issues and homophobia (see also Hubbard 2011a, Del Casino 2011). Multiple marginalisation was effected through both insults that are now outlawed by legislation and 'soupy' experiences (Yasmin, Chapter 4) that pervade all aspects of life but are difficult to name or are as yet little discussed. Some LGBT people found themselves pushed out on the basis of not fitting norms (including on the basis of class, gender and race). This resulted in forms of multiple marginalisations, where they were not in place in either LGBT or heteronormative spaces. Augmenting discussion of homonormativities, in Chapter 5, we noted that the pervasiveness of heteronormativities can be overlooked, and there may be neglect of the ways in which these affect LGBT people differently. Consequently, it is important to ensure hetero and homo normativities are critically explored in spite of, as well as because of, legislative equalities. Material issues of social and spatial justice were still pertinent for some, but not all, LGBT people in Brighton.

Nonetheless, multiply marginalised LGBT people also spoke of spaces they found or created where they felt included (see for example the creation of bi spaces in Chapter 5). There were also moments of feeling part of the gay city, even where prejudices had been experienced. For instance, in Chapter 4, we discussed participants' experiences where money and classed prejudice were clearly factors in using scene spaces, and John and Ed praised the scene and saw themselves as part of this aspect of the city. In this way, marginalised LGBT people were located within but pushed towards, and mindful of, the edge. This demonstrates the precarious nature of inclusion, as well as diversities between LGBT people. It moves between 'cynical resignation' and 'naïve optimism' (Halberstam 2011: 1), and asks for a recognition of the possibilities as well as normalisations of ordinariness.

Ordinary Possibilities: Becoming Commonplace

In examining the messy betweenness of those included and at times simultaneously excluded, we challenged the binaries that either celebrate lesbian and gay/LGBT legislative inclusion, or understand this as normalising (and therefore only privileging some and excluding others). It became clear in our participatory research that normalisation and regulation, *while* important, were also insufficient in conceptualising LGBT lives in Brighton in the first decade of the twenty-first century. Without negating exclusions and the forms of regulation imposed on those who have sought the position of normal (see Duggan 2002, Richardson 2004, 2005, Warner 1999), a sole focus on critiques can overlook social and political possibilities. Therefore, alongside the critiques of equalities gains, we *also* need to conceptualise the creative and cooperative possibilities of a time and place where that which was once legislatively and culturally deviant was rendered somewhat ordinary.

In order to develop considerations of these possibilities, we believe it is useful to distinguish ordinariness from normalisations/normativities. Normalisation can imply falling in line with a norm, an accepted way of being and living in a community. It relies on particular standards or norms, which here, only some can achieve or at least approximate. Inclusion through normalisations, such as that instigated through legislative measures, can mean that people are excluded in various ways, outside or at least on the edge, and marginalised. In contrast, ordinariness might not demand such compliance, where an infinite diversity of unspecified ways of being and living are regarded as unremarkable. In this way, anything can become ordinary through attitudinal change, rather than broadening a criteria of normalcy. Becoming ordinary therefore might refer to changes in how others are constituted, rather than looking at fitting or not fitting with norms or the notion of what is 'normal'.

The promise of ordinariness, of being unremarkable, was talked about as desirable and valued by CMIT participants, alongside accounts of enjoying being different. Engaging with how new inclusions were implemented in LGBT lives enabled explorations of the powerful move from being vilified and excluded, towards a position of being unremarkable. Some LGBT people, such as Andrew in Chapter 3, experienced being part of the normalised landscape where life situations, such as a being gay male co-parent, passed undistinguished in spite of differing from heteronormative orders. In Chapters 4 and 5 we argued that ordinariness could be created in microspaces such as scene spaces and through bi and trans groups. The separation of gay scene spaces, bi spaces and trans spaces from mainstream spaces could be viewed as challenging the notion of the ordinariness of LGBT lives. However, in Brighton, although these spaces were not the same as straight places, they were often understood as being part of the fabric of the city. Moreover, through acceptances in such spaces, some LGBT people became ordinary. Here, ordinariness was not reliant on replicating heteronorms. These spaces can create new norms and painful exclusions but they can also offer possibilities of ordinariness beyond normativities. Indeed they can create new conditions of ordinariness.

Instead of conforming then, ordinariness could offer a potential politics that asks to be unexceptional, but not necessarily adhere to homo/heteronorms. The power of being ordinary, which Weeks (2007) asks us to consider, lies in the possibilities of enabling inclusions without necessarily normalising. The challenge, therefore, is to create the possibilities of ordinariness that are not reliant on norms and otherings. One way of undertaking this is through the idea of ordinariness as commonplace.

Commonplace does more than allowing for LGBT people to be in-place, who once were out-of-place. It goes further to transcend this binary. When you are commonplace, you are not only 'in place', but common to the place itself. Place can then be shared or inhabited in common, as well as collectively created in ways that do not necessarily impose normative agendas. This returns us to place as a key constituent of ordinariness and its potentials. Becoming commonplace depends on the place (and time) where one might realise this. Ordinariness then is spatially contingent, just as critiques of homo/heteronormativities need to account for their spatial manifestations.

Our conceptualisation of ordinariness, through considerations of being or becoming commonplace, enables moves beyond the binaries of marginalisation/ inclusion, normalisation/queer. However, it does not exist in opposition to critiques of normativity. Rather, ordinariness can be strived for and sought *because* people continue to feel out of place in new sexual and gender landscapes and seek to be included, because of the failures of 'equalities' initiatives. The potential of commonplace is the possibility that anyone can become ordinary and this challenges the power relations that idealise particular normalised lives. We hope that this will begin discussions on a variety of directions for counter-posing the regulation and exclusions of normalisation.

Ordinary Activisms

Exploring the pursuit of a politics of ordinariness, in the second part of the book we addressed the possibilities of ordinary activisms in the first decade of the twenty-first century. These worked with and within 'the state' through partnership working, alongside other more oppositional and resistant activisms. This emerged in a very different era to when the state officially contested and blocked LGBT initiatives in institutionalised settings.[1] Where legislative equalities have been achieved, but marginalisations persist, questions arise about how LGBT political movements and activists might rework, reimagine and/or (re)strategise LGBT politics in contexts.

Activisms that work with and within potentially normalising processes and institutions may be part of the expected and accepted order. Some might contend that the response to these new conditions should be a critique of the normalising impulses of state provision and partnership working that seeks to cater for and include LGBT people, alongside other critiques of homonormativities. There can be little doubt that legislative and public service inclusions can, and should, be addressed through critiques of these normalisations. Throughout this book we have pointed to limitations, co-option and marginalisations, as well as the recuperation

---

1   Although it should be noted that during the late twentieth century specific local authorities and key individuals did seek to develop lesbian and gay work (see Cooper 1994, Cooper and Munro 2003).

of state power in working in and through partnerships with the community and voluntary sector.

In addressing these critiques, some have argued for a move towards radical politics that opposes new orders. The rise of queer activism is in part a response to what is seen as the mainstreaming of lesbian and gay politics (see Brown 2007). Creating and naming enemies through identifying new liberatory ideals and the disciplinary power of normativities produces an us/them binary, which can be empowering and contests dominant social norms. The conceptualisations of activisms that relate to the finding or creation of enemies to oppose are necessary for thinking through other possibilities and social worlds. Oppositional activist tactics that deploy the them/us, enemy/ally dichotomies are significant and in Chapter 7 these were important in addressing the recuperation of hegemonic power relations after breakdowns in partnership working.

Activisms that work with state providers in partnership might not be seen as 'political' by those who understand LGBT/queer politics as always oppositional and necessarily outside of the gates of power (Sears 2005). This was the case in Brighton where Spectrum selected tactics that trustees and workers believed would offer the best opportunity to create social change for marginalised LGBT people. These tactics were not seen as oppositional or 'shouty' in ways that for many defined what activism was supposed to be (see Chapter 6). While processes may strive to include LGBT people in the provision of public services and considerations of state policy, they are not *necessarily* or *only* normalising. Oppositional politics can be invoked in conjunction with other ways of doing, seeing and conceptualising political action. There is a need to engage with multiple possibilities of activisms, allowing for selection of strategies that have the potential to 'work' in each particular context, without negating the pitfalls of each. Perhaps then when considering the amelioration of LGBT lives, making sexual and gender diversities liveable, politics can take diverse and complex forms.

In this book, rather than solely critiquing the normativities of embedding LGBT equalities work in local government and public sector services, we *also* attended to the potentials that this form of activism enabled in places such as gay Brighton. Drawing on an understanding of 'the state' as messily recreated through our everyday lives in ways that contest the boundaries of state/non-state (see Andrucki and Elder 2008, Painter 2006, see Chapter 6), we moved beyond the binary of colluding/not colluding. When there is no monolithic state to collude with, there are a plethora of activist possibilities through working with, as well as contesting, partnership working.

There is a need to be territorially sensitive in considering social, economic and political contexts of pursuing sexual and gender liberation agendas. The messiness of state enactments operate in ways specific to their spatial and temporal contexts, thus so should the activisms that work towards sexual and gender liberations. Such a conceptualisation challenges generalised critiques of assimilation and co-option. In considering the messiness of the state and spatial specificities of Brighton, these activisms operated at the intersections of radical interventions and assimilationist impulses.

In the first decade of the twenty-first century in Brighton LGBT activisms included: working beyond normative/good gays towards 'queer' agendas; working with (and as) the state to provide public services for those who needed various forms of support; and working within powerful LGBT organisations such as Pride to attempt to make these cater for diverse LGBT populations. Enabling a multiplicity of activisms that are examined through their own territorial contexts, it becomes possible to hold in tension those politics that contest the state and those that include LGBT lives. For instance, in Chapter 8 discussions of Brighton Pride asked for a reconsideration of multiple possibilities of diverse (and imperfect) sexual and gender politics, including those that sought to celebrate LGBT people, contesting the sole representation of LGBT lives as despairing. Activisms that strive to make those who are out-of-place ordinary are different to assimilationist impulses that work to attain normative versions of hetero/homonormativity.

LGBT activisms that desire to be commonplace, rather than exceptional, can be seen as seeking ordinariness, rather than normalisations. Conceptualising ordinary activisms in this way does not neglect the need to critique continuing hierarchical power relations and normalising regimes that operate in inequitable ways (or political activisms that work to contest these). Indeed in distinguishing normalisation from ordinariness, each can speak to the other. This might entail considering how ordinariness might perpetuate disciplinary regimes. Conversely, it allows us to purport that becoming commonplace does not only involve normalisations and exclusions. This further challenges the ubiquity of solely critical thinking that Sedgwick (2003) has problematised and Halberstam's (2011) 'cynical resignation'. In this research, while there was a need to explore the new normalisations of partnership working and its limitations and failures in LGBT contexts, as well as more broadly (see for example Craig and Taylor 2003, Grundy and Smith 2007), there was also a need to value the political activisms that enable LGBT people to become ordinary, but not necessarily normative.

There is a danger of setting up ordinariness alongside queer transgressions and, in this way, creating a new binary of ordinariness/queer. We are not suggesting that there is one way of undertaking activisms and we see a need to challenge hierarchies of activist epistemologies, methods and practices that can solely valorise actions that seemingly stand outside norms and institutions. Robinson (2002, 2006) questions the distinctions and hierarchies between 'world cities' and 'ordinary cities' as politically dubious and developmentalist. She (2006) contends that we should avoid prescriptive directions for all cities (or, we argue, activisms), based on theorisations in a few. Instead of predetermining what activism is or should be, the focus can be turned to activisms in each specific context and their possibilities and problems. Taking territories seriously means *not* hierarchising particular activisms or reading sexual/gender activisms only through 'global' theories developed in specific metropolitan contexts (see Brown 2012). Instead accounts of activisms should be attentive to specific histories and geographies of the areas, which they emerge from and engage with. All activisms are contextually situated, and place is part of the social relations that provide the impetus and resources for action, as well as the means through which social control, hierarchies and inequalities are practiced. This allows for recognising that all activisms are partial and flawed, as well as potentially useful and effective, depending on their territorial settings. Therefore, we are calling for explorations of activisms that pay attention to the constitutive relation of space, moving beyond hierarchies and binaries.

What Next?

Jackie: I hope LGBT Brighton and Hove will regroup and re-radicalise when the Tories get in properly in 2011/12.

I hope that LGBT Brighton and Hove will remain diverse.

I hope that in the absence of statutory funding for most LGBT community sector services, which is coming, the private sector will step in.

I hope that central government will start to fund more local activism and particularly local project work around problematic areas like older LGBT, so that even if there isn't the local political will, on the basis of the existing laws there is some body that will help to enforce them and support any equality community that is experiencing institutionalised exclusion.

I'm hoping that Count Me In Too and its successor will make a huge difference to people's expectations of statutory services, even if there isn't the political will to follow through some of the changes that have begun in statutory services.

I'm hoping that what will happen is that the increased level of transparency and the increased level of evidence that is available to community members will provoke them to hold statutory services and private sector services to account and community and voluntary sector services that aren't as inclusive as they might be.

I'm hoping to be part of some really good work around bi identity because I think where there's a positive charge to being gay, and I don't mean positive in the sense of psychologically positive, I mean in the sort of literal physical sense, that there's a positive charge to gay identity or to lesbian identity and increasingly trans identity, that there's a discourse and there's a feeling that it's more than just a behaviour, that isn't there for bi people.

I'm hoping as well for a diversification in LGBT politics. That Stonewall will no longer be the single voice of lesbian and gay and bisexual people. I'm hoping that Stonewall will become trans inclusive and stop trying to gobble up local LGBT movements and start to admit that there are some issues other than its own that are important.

I'm hoping that there will be an increasing alliance between academics and the third sector so that there's a less scary academic discourse on the one hand and a more informed questioning way of being around identity, behaviour, social space etc., and equality hierarchy, all those issues, power, on the other.

I'm really, really hoping the Tories don't get in. I'm really hoping people don't fall for that because it will make a huge difference for the Equality Bill. It'll make a huge difference to the town. It'll make a huge difference to the community.

In her interview Jackie outlined her hopes, fears and speculations for the coming years. Her understandings were rooted and informed by her participation in public sector work and activism at the time of, and with, the CMIT research. Jackie's eloquent narrative is useful as it points to a range of areas through the differences between the 'now' when this interview took place (2009) and an imagined future beyond 2010. We chose to write about the lives and activisms in gay Brighton in the past tense because our research explores a specific time: 2000 to 2010. While the issues, concerns and problems highlighted in the research may well persist, contemporary understandings are likely to differ from those at the time of CMIT. This is in part associated with the significant changes in the national and local government context that Jackie feared would make a 'huge difference'.

In addition to historical contingencies, we have contended that lives and activisms are created through place. Taking into account that the analysis and learning we have presented is spatially and territorially sensitive, we are wary of offering universalising 'ways forward' for research. Instead academic research can extend understandings of the social and cultural contexts in ways that can inform effective activisms.

Jackie wanted to see increasing alliances between academics and the 'third sector' (community and voluntary sector). Examining and documenting new forms of LGBT lives and activisms in Brighton informed thinking and action in the city. In addition to deploying empirical evidence to address areas of need and make resource claims, activists can use scholarly thinking to engage with nuanced understandings of the multiple and complex issues that face LGBT people, what Jackie terms 'informed questioning'.

The processes of gathering material, writing and sharing ideas and thinking through this research project also constituted a form of activism. This book, written by, with and for activists, as well as academics, continues this process. Nonetheless, it comprises only a fraction of the possible stories that could be told regarding new forms of LGBT lives. In addition there is much more to be said about gay Brighton, and LGBT lives therein, than can be encompassed in the pages of this book. We hope that this book will more broadly spur considerations of the positive effects of new political landscapes, alongside the critiques of normativities that have so ably been developed. Considering these in spatially sensitive ways opens horizons for valuing a plethora of activisms and hopeful possibilities where they occur.

Appendix 1

Count Me In Too: Partnership and Process

This technical appendix gives details of the methods used in this research. It gives the background material for the entire book but particularly Chapter 2. This simplifies what was a complex project for the purpose of introducing the reader to the research processes.[1] We begin by outlining the partnerships that were crucial to the Count Me In Too (CMIT) research. Developing our discussion of how community/voluntary, statutory/public and university sector partners negotiated the research process, we discuss the methods used in identifying priorities, data collection, data analysis and dissemination of the CMIT research.

Partnerships: Working Between Community, Statutory and University Sectors

In CMIT we sought to disrupt research models in which policy makers and researchers (academic or otherwise) generate knowledge regarding marginalised communities without their input (mrs kinpainsby 2008), or where research is undertaken without input from service providers and policy makers. Here we worked across public/statutory, community/voluntary and university sectors, challenging the boundaries that limit initiatives to working across two of these three areas. There was significant and deliberate reworking of the assumed hierarchies that place public services and Universities in positions of power over communities. At times, Spectrum and the University of Brighton researchers operated within the traditional boundaries of academic and community partner. However, this divide was more often blurred, contested and challenged. The roles each played in this research are outlined in Appendix Table 1.

Rather than being owned by services and framed as 'community engagement', CMIT involved a range of stakeholders as equal partners in a project owned and managed by LGBT people. This is not to deny that various forms of power were in operation. Those involved in this research process were subject to a matrix of power inequalities that impacted on partnership working. The specificities of partnership working for this research related to the use of a research team who interacted with public and statutory services and sought to empower LGBT people to create and own the research (see Chapter 6 for a critical discussion of partnership working).

---

1 Further details can be found at www.countmeintoo.co.uk, in the 'community summary of process' and throughout the findings reports.

Appendix Table 1 Simplified summary of partner roles in Count Me In Too

Sector	Role in CMIT
Public/statutory services	Funding, submitting questionnaire questions, analysis groups, actioning many of the recommendations; attending feedback meetings and co-producing press releases.
Community/voluntary	'Holding' and 'owning' the process- ensuring diverse engagement across LGBT people, questioning and challenging funders and researchers; sitting on steering group, action group, analyses group; submitting and choosing questionnaire questions; drafting focus group questions; actioning certain findings; writing certain summaries and co-authoring certain reports and academic outputs.
University	Research skills and expertise; facilitating key groups; data collection and analyses; funding; reputation of 'academic' research; offering validity and 'weight' to the findings; drafting and writing all reports and most summaries; authoring academic outputs.

The research team consisted of academic and Spectrum researchers. It devised and managed the research and engagement processes, publicising 'news' from CMIT and opportunities to join the process, thus inviting participation, managing expectations and moderating actions of those who took part. The research team held power managing how the resources of the project were used, or even, consumed: the time, energy and involvement of those who took part, openings for engagements with external organisations, as well as funding. While decisions about the design of research, the strategies for data collection and the analyses were made with involvement of those consulted, the processes of the project at each stage were decided 'behind closed doors' by the research team and a stronghold of power potentially sat with the academic researcher (Kath) who advised on method and analyses. However, we sought processes and attitudes to disperse power and negotiate difficulties that arose from power imbalances. At a pragmatic level, the decision-making delegated to LGBT people was practically focused, shaping what happened next, and didn't require long term involvement. However, the operation of the research team also set up somewhat 'impenetrable' process. The research team used power in order to ensure the project could be established and progress, and there was a series of negotiations in the operationalisation of the research. We now turn to the specifics of the methods and in doing so outline some of these negotiations, through how this research worked in partnership across community, statutory and university sectors.

Operationalising Community-Led Approaches and Identifying Research Priorities

The project was directed by a community-led approach using volunteers from LGBT communities who made up a *steering group* working on project design and an *action group* engaging in data analysis. From the outset, the research team sought to engage a range of people who identified with, and potentially offered contacts in, more marginal sectors of the LGBT communities. Individuals, statutory services and community and voluntary sector groups were invited to identify priority issues and themes for research, to interpret findings and evolve responses at five stakeholder meetings which were held at key points in the project.

Initial discussions concluded that Spectrum as an organisation represented by a worker and trustees could not represent the diversity of the Brighton and Hove LGBT communities. The *steering group* was made up of LGBT people who in some way identified as multiply marginalised (see Appendix 2). In this way we sought a means of involving and hearing from people with a range of identities and perspectives. These were identified through Spectrum's existing contacts and people also volunteered in response to Spectrum or CMIT events and press releases. However, it became clear that people who felt disempowered, disengaged and excluded were unlikely to feel confident in volunteering. Therefore engaging individuals in discussions that highlighted the relevance and importance of their skills and experiences was how many of those who formed the steering group were recruited. Because of this, the steering group was made up of those who in some way felt part of, (and therefore attended a meeting or got in touch in some way with the research), and simultaneously felt marginalised from, LGBT collectives. The specialist knowledge of steering group members was deployed to:

1. Identify 30 factors that marginalised LGBT people. This list informed who was invited to take part in focus groups (see Appendix Table 3).
2. Design the questions for the focus groups.
3. Make decisions regarding the structure and issues examined in the questionnaire.
4. Recruit participants to focus groups.
5. Run a pilot focus group made up of steering group members who were able to comment on the process as well as the questions.
6. Promote the questionnaire.

Decision making by steering groups places influence in shaping the project with those being researched, but was of course mediated by the researchers (see also Browne, Bakshi and Law 2009). We now outline the data collection methods that were designed by the steering group.

Data Collection

The data collection for CMIT involved a series of focus groups, a questionnaire, consultation and feedback events and key interviews. We describe these methods in turn, detailing key aspects of each.

Focus Groups

The focus groups ran from February 2006 to December 2006 and their aim was to explore LGBT voices and needs that are often not included in large-scale questionnaires. Individuals were recruited to the groups using the CMIT and Spectrum websites, along with publicity in local gay and mainstream newspapers, and word of mouth and personal contacts were important in recruitment. There were 19 focus groups and one interview[2] with a total of 69 participants (see Appendix Table 3). These lasted between one hour and four hours. The focus groups followed the headings designed by the steering group (Brighton and Hove; Being Excluded/ Discriminated Against/Left Out/Made to Feel Like You Don't 'Fit'; Safety; Health and Wellbeing; Support and Services; Housing and Relationships;[3] Actions and Priorities for Change). These sections mapped onto questionnaire headings and enabled deeper examination of complex issues by drawing on personal experiences of multiple marginalisation, using open qualitative questions.

The list of multiply marginalised identities that was drafted by the steering group served as a guide in convening focus groups (see Appendix Table 2). However, it was not possible to undertake focus groups with the full range of identities that were identified (see Appendix Table 3). In order to ensure that everyone who wanted to participate in a focus group could do so, two 'general' focus groups were held for those who had missed a particular group, or who were the only person who came forward in their identity category. Individuals' multiply marginalised identities were acknowledged by inviting participants to

---

2   They were: Bisexual, Black and Minority Ethnic, Deaf, Disabled People, Domestic Violence, Older People, First Generation Immigrants, General Focus Group, Hate Crime Survivors, Mental Health, Pride in Whitehawk/those living in an outlying estate), Parents, Pilot, Trans, Women and Young People.

3   Due to the sensitive (and assumed individual and private) nature of housing and relationships questions, it was deemed inappropriate to address these issues in focus groups. A confidential questionnaire about these issues was designed specifically for completion by focus group participants after the group discussion had finished. Questions were mainly open ended to encourage in-depth answers. These questionnaires were handed out to participants in all the focus groups with a stamped self-addressed envelope. They could be completed after the focus group by hand or by verbal interview with the researcher. No one chose to complete the questionnaire by interview after the focus group, although some did fill out the questionnaire before leaving the focus group venue and handed it to the researcher. Seventeen of these additional questionnaires were returned out of the 68 participants.

attend any focus group with which they identified, regardless of their attendance at other groups. This ensured that participants could discuss the breadth of their experiences in appropriate and supportive environments.

Many who participated in the research said that they appreciated focus groups that acknowledged multiple and intersecting identities. Some people told researchers that it was the first time they had shared experiences with people 'like them'.

Appendix Table 2 Multiply marginalised identities listed by the steering group during the process of designing data collection methods

Young people	People on low incomes
Older people	People with mental health difficulties
Disabled people	People on isolated estates
LGBT Parents	People in prison + ex-prisoners
Deaf people	People in supported accommodation
Trans people	People in residential / institutional care
Bi people	Asylum seekers / refugees
Sex workers	Substance + alcohol users
Overseas students	People with learning difficulties
Carers	People of faith
Homeless people	Survivors of domestic violence
BME people	Women who don't access mixed spaces
Celibate people	Men who have sex with men
Travellers	People coming out later in life
Hate crime victims	New to Brighton

Appendix Table 3 Focus group details

Focus group	Number of participants
Bisexual	7
Black and Minority Ethnic group 1	3
Black and Minority Ethnic group 2	2
Deaf	2
Disabled	5
Domestic Violence Survivors	1 (interview with support)
Elderly	4
First generation immigrant	3
General focus group 1	4
General focus group 2	2
Hate Crime Survivors	6
Mental Health	2
People living on an outlying estate	4
Parents	2
Pilot	3
Trans group 1	4
Trans group 2	3
Women's	2
Young People group 1	6
Young People group 2	3

The Questionnaire

The CMIT questionnaire aimed to identify LGBT needs in Brighton and Hove. It targeted all LGBT people who lived, worked and socialised in the city. The process of designing the questionnaire illustrates a key aspect of the project's approach to partnership working that sought to empower marginalised LGBT people over statutory service providers. LGBT people in the steering group decided the direction of the research and what was to be included in the questionnaire.

Between December 2005 and January 2006, individuals, community groups and statutory services were invited to submit questions that they could use in their work with LGBT people. Over 400 questions were submitted. These included questions from national surveys, used to establish how national findings compared with local LGBT people's experiences, as well as questions exploring local issues that were believed to be having an effect on LGBT people in the city, but may not be recognised nationally.

The importance given to statistical data was apparent in the questions submitted. Alongside these, qualitative questions gathered further information on experiences and opinions. The academic researcher and the Spectrum worker identified gaps in the questionnaire, particularly around multiply marginalised identities as categorised by the steering group. Further stakeholders were then contacted who were able to advise on specific sections and questions that would be of use to service planners and providers, including LGBT services. The steering group reviewed all the questions submitted, reduced the number of questions and edited them where appropriate. Finally the questionnaire was checked by academics and the person in charge of setting up the online questionnaire system for consistency and clarity and to ensure the question phrasing allowed for reliable data to be collected.

The resulting questionnaire had 238 questions with a series of routings around specific identities and experiences, which allowed different experiences and identities to be explored in more depth using quantitative and qualitative questions (the paper version of the questionnaire can be viewed at www.countmeintoo. co.uk). In April 2006 the questionnaire was previewed with stakeholders (all those who had submitted questions and other interested individuals). A stakeholder meeting was held with the aim of ensuring that there were realistic expectations of the project and to gather feedback on the process so far.

The questionnaire ran from May 2006 until October 2006. There was an online version and a paper version to increase access for a range of LGBT people. In addition outreach services supported clients to complete the questionnaire. There was also the opportunity to seek help from University of Brighton staff for anyone who had difficulty completing the questionnaire and free access to the online questionnaire was given at a local LGBT cafe. The questionnaire was advertised through word of mouth, email communications, flyers and posters, the CMIT website, newspaper articles and advertising, radio interviews and links from other websites.

A total of 819 valid responses were received from the questionnaire. Some key demographics (aside from those outlined in Appendix 2) were:

- 56 per cent of the sample was male, 41 per cent female with 3 per cent defined as other gender categories.
- 60 per cent of respondents were aged between 26 and 45, 15 per cent were under 26 and 10 per cent over 55.

- Just under three per cent of the sample identified as BME and just over three per cent identified as an 'other' (i.e. non-white and not BME) ethnicity category.[4]
- 15 per cent of the sample had a disability. Three per cent of the sample identified as being deaf, deafened, hearing impaired or deaf blind. Seven per cent of the sample had tested positive for HIV.
- 16 per cent were parents or closely related to young children.
- 10 per cent of the sample earned under £10,000, with 12 per cent earning over £40,000.
  (see Browne 2007 for a full breakdown of the sample characteristics).

Consultation Events

A consultation event and a final feedback event also provided data that is used in this book.

The first consultation event took place in April 2009 following publication of the report and securing of additional funding for dissemination. This was a community consultation event that sought input from a variety of stakeholders regarding the next phase of the project. This event included:

- a poster exhibition about the project to date,
- several mechanisms to collect comments to inform the broader dissemination strategy,
- a speakers' corner at which LGBT people linked to the project were invited to speak for five to ten minutes on CMIT. Contributions ranged in scope and content and individual speakers addressed different issues. Twelve speakers were audio recorded and transcribed. These recordings form data that is used in this book.

The second event was a final feedback group where the research team (consisting of the University and Spectrum partners in the research) and all involved in CMIT were invited to discuss comments on CMIT and considerations 'for anyone doing Count Me In Three'. This took place at the close of the project and enabled reflection on the project as a whole, as well as on the first decade of the twenty-first century. Seven people took part in group discussion at this event.

Stakeholder Interviews

LGBT Stakeholder interviews were undertaken later in the project (2009–2010). They were designed to provide a counter balance to the project's focus on speaking

---

4   BME was used instead of the census ethnicity categories as this is what was asked for by BME members of the steering group. They felt that the census categories were ill defined and that they often felt they occupied multiple 'boxes' but were confined to one.

with marginalised LGBT people. To this end they targeted LGBT individuals who were identified as 'holding power' in some way in 'gay Brighton' or in key statutory services. For the most part we were interested in those who had a history of involvement in LGBT issues in the city.[5] However, as became clear throughout the interviews, there was at times no clear division between those who experienced forms of multiple marginalisation and those who held key roles in the city.

These interviews lasted between one and five hours and 19 people took part. The interview questions drew on the data collected, analysed and reported in CMIT, as well as addressing the gaps of the research. All were familiar with aspects of the CMIT project through being part of the project and/or working with the project findings, partially because of their positions in the city. The stakeholder interviews gave an insight into processes and relationships that were not captured in discussions of LGBT lives in the questionnaire and focus group data.

Ethical Procedures

Ethical considerations were also central to the research process. The questionnaire questions were reviewed by experts in the area, particularly where these were sensitive, for example, regarding domestic violence. Where they existed, helplines were listed on the countmeintoo.co.uk website. However, it was noted by the steering group that this could be disempowering where groups did not exist and expectations of support and help were not met (see for example Chapter 6). The option of contacting the researchers and the community partners was given on the website and on the questionnaire. At each focus group, clear guidelines were given regarding confidentiality and other matters, originally drafted by the steering group, and further guidelines were invited. Each participant was given an information sheet and a consent form with the option of withdrawal presented at the start and the end of the group. All participants were offered the option to receive their contributions recorded in transcript (not the entire transcript) and a brief outline of the main points of the focus group in which they participated.

There was a further ethical and moral responsibility on both Kath Browne and Spectrum to ensure that the data was used for the purposes intended and not misused by individuals, groups, business, services or the media. Consequently, the questionnaire databases and primary data was owned and managed by Kath Browne and Spectrum. This included all reports, summaries and dissemination and both parties saw this control as necessary to ensure that this trust given by those who were involved in this research was protected. Since the demise of Spectrum (see Chapter 8), this responsibility lies with the Count Me In Too Research team, and particularly Kath Browne and Leela Bakshi. It is hoped that after a suitable

---

5   It was not possible to interview all those who were identified as fitting this criteria, and we encourage further work to document and record the histories of 'gay Brighton' in this way.

period that the Count Me In Too data and other archival material can be archived with an appropriate LGBT history project.

For this book pseudonyms are used for all focus group respondents and for those in feedback and consultation meetings. Questionnaires are referred to by number.

Interviewees were given the option of being named and, where they requested to be named, their first names are used. For those who did not want to be named, and because of the identifiability of interviewees, multiple pseudonyms are used in order to further safeguard identities. Groups and services referenced are also removed where interviewees could be identified through them.

After showing their quotes to some of the interviewees, they requested they were tidied up for presentation on the page. They felt that alongside the academic text they came across as inarticulate and 'stupid'. This was not our intention and in respecting our participants, we have taken on-board this request and applied it to all focus group and interview quotes. This is because, in the main, the points are more important than how they were said. Retaining what was said means that at times we use lengthy quotes, allowing participants' transcribed voices to make complex and nuanced points.

Data Analysis

Data analysis took various forms throughout the project, reflecting its evolving processes. We group these into three rounds of analysis: the initial findings, further analyses and analysis for this book.

The first round of analysis was undertaken by an action group consisting of LGBT people with the capacity to trawl through reams of data and offer insights into the priorities for the initial findings.[6] The action group met to analyse the findings over a number of Saturdays between February and June 2007, and were given data to read in preparation for these meetings, with the aim of beginning the initial analysis of the data and identifying key findings for the initial report and dissemination event. The focus group data was collated into key points, anonymised and passed to the action group to read. The percentages from the questionnaire were also given to the action group and, for a large proportion of the questions, the action group requested that differences between the LGBT communities be explored. Therefore questions were cross-tabulated on the basis of 10 identities,

---

6   The action group was conceptualised by one of the steering group, who applied for funding for meetings from Brighton and Hove City Council for room hire and refreshments. A stakeholder event was also held in December 2006 and attended by representatives from a wide range of groups and organisations that were updated on processes and timescales for the analysis phase of the project, and invited to submit questions to be examined in the data analysis.

viewed by the action group as significant (age, deafness, disability, ethnicity, HIV status, isolation, income, mental health, sexual identity and trans status).[7]

In-depth analyses of the data continued throughout 2007–2008. During this time eight further analysis groups explored the data with regard to eight themes, informed by priorities in the city and nationally. These themes were domestic violence and abuse, safety, housing, mental health, health, trans people, bi people and drugs and alcohol (see www.countmeintoo.co.uk for full reports and community summaries). The analyses groups for the further detailed reports were made up of representatives from a range of statutory services, voluntary groups (these representatives included those who identified as LGBT) and LGBT people who operated in an independent capacity. All analysis groups were closed and confidential, with data only released at findings events. In further analysis groups, workers from statutory services and LGBT individuals took part with equal status, working with researchers to identify key findings. They suggested cross-tabulations and other further statistical analyses; explored anonymised quotes from focus groups and questionnaires; suggested interpretations; and pointed out implications of the research in working towards improving LGBT lives in Brighton and Hove. Finally, analysis groups suggested (and owned) recommendations relating directly to strategic priorities and specific programmes. These were checked by Kath, as the academic on the project, to ensure that the recommendations were substantiated by the data in the report. However, it was beyond the academic role to identify strategic priorities and strategies. Recommendations sought to redress some of the troubling findings in the data by suggesting strategic initiatives relevant to particular themes, mainly calling for training of front line staff, better engagement with LGBT people and respecting differences between LGBT people.

All of the data was reanalysed for this book. This included coding and analysing the interview data and revisiting the questionnaire and focus group data. The key themes were chosen from the data, but also related to academic questions and areas that could offer insights into LGBT lives and activisms. In the main the data was re-analysed by Kath, who led all of the analysis groups and used these priorities to guide the analysis that led to this book. The key themes were then discussed with Leela and the analysis developed through the writing process (as discussed in the Preface).

Dissemination

The project presented its findings in a variety of ways designed to communicate the data and findings to a broad range of audiences, in a variety of formats. Dissemination activities also sought to influence policy agendas and service

---

7   The questionnaire data was analysed in SPSS (statistical package for social scientists) and tested for statistical significance using chi-squared and t-tests. The significance level used is $p<.05$.

provision for LGBT people (see www.countmeintoo.co.uk for reports, summaries, conference presentations and other materials). The key elements of the dissemination were:

- *Initial and community reports.* The initial report was written by Kath with the action group advising on the data analysis. It reported key results from the research as defined by the action group. Members of the action group then constructed a community report that offered easier read key points from the findings and included some interpretation of the findings.

- *In-depth reports and summaries* from the eight further analyses groups.

- *Launch events.* These events brought together service providers, practitioners, partner groups, organisations and members of LGBT communities, at events hosted by Spectrum. These events sought to influence social change by engaging with service providers, and findings events were designed to be beginnings rather than endings. The presentation of the research took up less than half of the session. The second half of the event consisted of an extended period of questions and answers with a panel of experts including representatives of services (often attended by heads of services), commissioners and members of the LGBT communities and groups.

- *A conference entitled Transforming LGBT Lives: Research and Activism for Gender and Sexual Liberation* (September 2010). This brought together academics, public sectors and LGBT communities, and invited papers and reflection on future work in a changing political and economic climate. It also addressed issues that were not covered in the findings, such as LGBT carers.

- *A travelling exhibition of research findings.* This was created in response to calls at the consultation event in the April 2009 to communicate the findings to mainstream gay people and heteronormative people. For the former, the exhibition was displayed at events, such as the Golden Handbags Awards,[8] where posters were shown and key findings communicated through flyers and displayed on the main screens in between acts. A week-long exhibition of findings posters was part of the 2009 Pride in Brighton and Hove festival. This culminated in a CMIT tent at the main park event. The exhibition was also taken to the well-used central Brighton library in Jubilee Square, during LGBT history month in February 2010. The University of Brighton

---

8   This is a gala awards ceremony hosted by the local gay magazine where awardees are voted for by 'the public'. Categories of nominations include best bar boy/bar girl, best club, best bar, best manager, best door security person. There is one category for the voluntary sector, but an award is also given by the editor of the local gay magazine to people or groups he feels deserve them. The event attracts over 500 people from the gay businesses across the city.

Community University Partnership Programme events and conferences also displayed the CMIT findings.

People left feedback for the project and reflected on the impact of Count Me In Too using forms provided at these events. This evidence demonstrated that collectively, these exhibitions raised awareness of contemporary LGBT issues in Brighton.

- *Presentations of findings* took place at various events and to committees nationally, internationally and across the city. These were presented by the research team (mainly Kath Browne, Leela Bakshi, Nick McGlynn and Sharif Mowlabocus) and included keynote talks at a variety of academic and service provider conferences.
- *Academic papers* in peer reviewed journals aimed to share learning from the project with wider academic and activist communities, and also to 'validate' the research in contexts where 'peer review' is seen as the 'gold standard' of knowledge (see Chapter 2).
- *A website* (www.countmeintoo.co.uk) was created containing all the resources from the project, news reports, research team contacts, presentations and a video from the final conference, and a video from the 'wish tree 'which remain accessible online. This is a lasting resource now that the project has finished.

Appendix 2
Defining Lesbian, Gay, Bisexual, Trans and Multiple Marginalisation Used in This Book

Introduction

Sexual and gender identities are fluid, diverse, complex and spatially and temporally created (Brown 2000, Bell 2001, Binnie 1997, 2004, Browne et al. 2007, Knopp 2004). Terms in use might not encompass all who may be recognised through categories such as Lesbian, Gay, Bisexual and Trans (LGBT), indeed these labels may be actively resisted by some and have been challenged through discussions of intersectionalities. Yet they were deployed in the Count Me In Too (CMIT) research and we use them throughout this book. This technical appendix provides an overview of the reasons for our use of specific terminology in relation to the poststructural conceptual basis we use. It addresses the issues that can arise at the interface of queer thinking and social science research (Browne and Nash 2010). We examine the import of deploying and using these categories in geographically sensitive ways, firstly justifying our use of the term LGBT. We then discuss the use of the terms bi and trans, and finally, we outline our understanding of intersectionalities and our use of the term multiple marginalisation.

Uses of LGBT identities

Contemporary challenges to LGBT identities question the usefulness of these categories that exist only because they are used. Arguably this undermines the premise of the book which works with and across the category LGBT. We use this part of the appendix to deal with this potential incongruity. Queer[1] thinking emerging from the humanities can challenge the foundationalisms of identity-based categories. Critiques have contended that by solidifying behaviours, relationships, practices and attractions through categories such as LGBT, what they name is reproduced and normalised, rather than mirroring a pre-existing reality (Jagose 1996). Such categories then hierarchise, limit, restrict and homogenise individuals and their practices. Moreover, the increasing normalisation of

---

1 We use queer in relation to queer theory, rather than as an umbrella collective for LGBTQI (lesbian, gay, bisexual, trans, queer and intersex), denoting a methodology for exploring non-normativities, that are abnormal and defiant (see Browne 2006a, Giffney 2004, Giffney and O'Rourke 2008, Oswin 2008, Ward 2008).

terms like lesbian and gay have brought LGBT identities under further scrutiny questioning their transgressiveness, through investigations of homonormativity and homonationalism (see Chapter 1, Duggan 2002, Puar 2007). In contrast, empirical studies of the identity-based category LGBT start from the premise that these categories are meaningful and important, and that certain identity groupings are often socially excluded in various ways. While these positions are over-simplified caricatures of the diverse fields of sexuality, gender studies and queer thinking, they offer useful starting points in thinking through the contestations regarding LGBT identities.

Identification within the categories of trans and/or bi, lesbian or gay is complex and fractured, and the 'rainbow' of identities can call into question the categories themselves (Browne and Lim 2010, Lim and Browne 2009; see also Monro 2003, Johnson 2007, Stryker 2006, Valentine 2007, Whittle and Lewis 2007, Roen 2001). Notwithstanding that LGBT identities are recently deployed (Weeks 1989), they can form the basis of equalities, rights claims, legitimisation and political action. The complexities and the fluidities of categories can be politically important (see Halberstam 1998, Valentine 2007). In asking academics to be clear about who we include in the category transgender, Valentine argues that this category should remain flexible in order to maintain its 'strength as a tool of political organising' (2007: 39). LGBT identities can also allow for belonging, recognition and alliances, as well as being able to recognise marginalisations. There is power in naming and this can be used to address oppression, marginalisations and exclusions (Crenshaw 1995, in Erel et al. 2011). Thus, categorisation per se is not necessarily disempowering (or limiting, see Green 2010).

We mobilise LGBT categories for political purposes, recognising that there was significant heterogeneity in how people in the CMIT research defined themselves. We use the term/s lesbian, gay, bi[2] and trans,[3] because they were politically expedient and spatially deployed in ways that had the potential to ameliorate the lives of those marginalised by their gender and sexual identities. We do this recognising that these terms are complex, multiply appropriate, refuted and questioned. Nonetheless, activists and others in Brighton wanted the terms LGBT generally, and bi and trans specifically, to be recognised by public sector and LGBT services and others (see Lim and Browne 2009, Voss et al. 2012). They formed useful and important ways of gaining recognition and also stood as a basis to make rights claims, as well as justifying allocations of resources, such as for mental health provision and gender reassignment (see Browne and Lim 2010, Browne and Bakshi 2011, Lim and Browne 2009, West 2004). In drawing on these categories as recognisable and readily used terms in the context of sexual

---

2   We use bisexual at times for clarity, where it was used by participants and where questionnaire data refers to bisexuality.

3   In order to discuss the differences we use the terms 'trans people', 'trans respondents' and so on, and contrast this with 'non-trans', which we use interchangeably with the term 'cisgendered'.

identities in the UK, we recognise that strategic deployment of these categories can effect erasure and marginalise groupings, such as bi and trans people (as well as those who remain unnamed and unrecognisable, Butler 2004). Thus, we sought a conceptualisation which allowed for differences to be seen, without negating the possibilities of deploying the umbrella of 'LGBT'. We now move to look at the use of the terms bi and trans specifically, before addressing the possibilities of alliances and uses of the overarching term LGBT.

Bi

Bi as an aspect of LGBT is often occluded in discussions of gender and sexual identity, or presumed to be catered for by queer, gay or straight discussions (Angelides 2006, Gammon and Isgro 2006, Gupta et al. 2010, Hemmings 2002). Barker and Langridge (2008) name bisexuality as a 'silenced sexuality', and the erasure of bi identities is a key issue for bi people. This erasure impacts on communities, socializing, mental health, as well as on accessing services (see for example, Jorm et al. 2002, Oxley and Lucius 2000, Barker and Yockney 2004). Indeed, often homophobia and transphobia are the only discriminations mentioned in relation to LGBT people. This elides and erases the importance of biphobia as a discrete form of discrimination, oppression and exclusion. In addition, queer theory, which could be an ally in moving beyond dichotomous thinking, has been found wanting (see Barker and Laindridge 2008, Voss et al. forthcoming). Hemmings (2002) explores how bi histories are told and retold, as well as noting those that fail to be retold and are subsumed and forgotten. Thus, like Hemmings, this book stakes a claim for bi identities, resisting their erasure and dissolving into LG, straight and/or queer. Yet, the meaning of 'bi' and the theoretical understandings of this identity positioning are multiple and complex (see for example, Hemmings 2002, McLean 2007).

In Brighton, groups and individuals used bi identities in order to be named (see Erel et al. 2011). This enabled bi people to be seen, heard, differentiated from, as well as placed within, the grouping 'LGBT'. We use mainly use bi[4] throughout this book as this was the term preferred by our participants and those who created CMIT through steering groups, actions groups and analysis groups (see Chapter 3). Bi allows for identities, desires, politics and communities that are not only located in the 'sexual' but can always be related to it. This term is of course contested, fluid and constructed, however as Hemmings and others contend, it remains significant in creating spaces and 'middle grounds'.

---

4   We use bisexual at times for clarity, where it was used by participants and where questionnaire data refers to bisexuality.

Trans

We follow what Roen (2001) terms a 'liberal politics' in engaging with trans people, focusing on legal protections, the provision of services and the safety of trans people. This was the focus of aspects of the Count Me In Too research that used participatory research methodologies, informed by trans people who participated in framing the research. This is not to say identifications or politics were straightforward or readily agreed. We have shown elsewhere that, in the CMIT research, identification within the category 'trans' was complex and fractured, and this 'rainbow' of trans identities calls into question the category itself (Browne and Lim 2010, Lim and Browne 2009, see also Monro 2003, Johnson 2007, Stryker 2006, Valentine 2007). Yet, this category was useful and important both in terms of recognition for trans people and as a basis to make claims for services and provision (see Browne and Lim 2010, Lim and Browne 2009, West 2004). Therefore, we recognise that the category of trans, as with all gender and sexual identities including heterosexuality, is internally heterogeneous, and might not encompass all who may be recognised through it; it is actively resisted by some and is fluid and unstable (see Chapter 1, Namaste 2000, Johnson 2007, Stryker 2006, Valentine 2007). We use it here and in our research because for some, it enabled a voice both within and outside of the broader category of lesbian, gay, bisexual and trans. We recognise that this remains a contested point.

Various categories and umbrellas have been used to label those who we term 'trans' (see Browne and Lim 2010, Brown and Bakshi 2011). For us, the rejection of sex and gender distinctions, pushes us away from the use of transgender or transsexual, towards 'trans' as a category that neither privileges sex or gender, nor recognises the necessary separation of sex from gender (see Lim and Browne 2009, Browne and Lim 2010).[5]

Alliances: Working Across LGBT

It could be contended that given the differences between L, G, B and T people, these should be treated and addressed separately. Indeed there have been numerous important in-depth investigations of the distinctiveness and specificity of these identities, as well as the differentiations between categories within this grouping. However, undertaking studies that focus on specific aspects of LGBT collectives does not facilitate comparisons between these groupings and can also overlook how alliances can be built across these categories. For instance, separating T from LGB on the basis that one is about gender difference and the other sexual orientation can neglect shared histories and contemporary realities of multiple or same-sex desire that were and are performed, read and policed through gender deviance, in particular disrupting the hetero-gendered norms of male and female (consider for

---

5   Although it should be noted that the move away from 'fleshy' transsexual bodies is contested by some (see Coogan 2006).

example the ways in which 'homosexuality' was once understood through gender inversion and now is often associated with effeminacy, see Valentine 2007). Understanding that the collective category LGBT is geographically created, in certain contexts as solidarities, alliances and connections between LGBT people, can render this a viable category to both study and discuss (see also Chapter 5).

There are political effects of a lack of solidarity and segregation of trans and gender outlaws. In the US, the sponsor of the 2007 Equality Non-Discrimination Act withdrew gender identity to ensure the passage of the bill protecting employees on the basis of sexual orientation (Green 2010). In the UK, the passage of key legislation for trans people has been behind that for lesbians and gay men, and key organisations, such as Stonewall, that define themselves only as lesbian, gay and (at times) bisexual do not campaign for this. Understanding that not only trans, but also the collective category LGBT is geographically created, we have argued elsewhere that in Brighton, solidarities between LGBT people make this a viable category to both study and discuss (see Browne and Lim 2010, also Chapter 5).

Deploying LGBT in the Count Me In Too Data Collection and Dissemination

Although we see LGBT identities as fluid, conditional and difficult to pin down, we use them for the purposes of allowing for belonging, and recognising exclusions and marginalisations. Although categories can be formed through practices of inequality, there is power in naming and this can be used to address oppression, marginalisations and exclusions (Crenshaw 1995, in Erel et al. 2011). Working with the complexities of deploying LGBT identities and recognising their fluidities and contingencies, we included the following explanation in the FAQ's for the online questionnaire used for data collection:

> LGBT is a shortened version of lesbian, gay, bisexual and trans, however, we are using it much more broadly for ease of use of the questionnaire and also to encompass the broad range of diversities that this questionnaire seeks to include. This term then includes:
> * Men who have sex with men,
> * Those who are questioning or unsure of their sexuality or gender identity,
> * Those who define as heterosexual but have sex with, or are currently in a relationship with, one person or people who are the same sex to themselves,
> * Those who do not define their gender/sex/sexuality but live outside of heterosexual norms,
> * Womyn identified womyn,
> * Those who choose to queer/play with categories of gender, sex and sexuality.

We know that LGBT is not a perfect term but we have yet to find 'the term'. Queer is often used for this purpose. However, this term can be associated with bullying and socially and academically has many meanings.

We know this term hides multiple experiences and differences between us. However, if we want to address need and explore what can be done to make Brighton and Hove a better place to live, work, socialise and use services in, we need an overarching term. Throughout the questionnaire you will be given a chance to discuss your difference and sameness within and outside this term. This will hopefully enable you to express your needs, your identity and yourself!

In the initial report we also explained our use of the terms 'sexual identity' and 'trans':

Sexual identity, rather than orientation, is addressed in this research. This of course can exclude those who do not identity as lesbian, gay, bisexual, however, it avoids the dubious categorisation of behaviours and 'desires' inherent to assessing 'orientation'. Sexual identity presumes a self-awareness and understanding that is usually associated with processes of 'coming out'. The study did not preclude those who identified as heterosexual and trans, or heterosexual and had sex with members of the same sex.

Gender variance is also a difficult concept. The complexity of this is homogenised under the term trans, however, there is a multiplicity of identifications and experiences. These are not essentialised to body composition or transitional 'stages' in this research. Consequently trans is the preferred term rather than transgender or trans sexual (see West 2008, Whittle et al. 2007).

The term LGBT is used for ease and understandability. The author/s recognise the difficulties of categorising sexualities and gender identities in this way. The term includes those who are questioning, unsure and do not identify with particular sexual or gender identities.

Nonetheless, in utilising quantitative research methods, categorisation is inevitable (see Browne 2008, 2010). For example, there can be little doubt that, given the complexities of trans identifications, quantitatively categorising 'trans' and gender identities is also problematic.[6] The question in the CMIT survey read, 'Do you identify yourself as being trans or have you ever questioned your gender identity?' Five per cent of the sample said yes to this question and were defined as trans (n. 43, with 92 per cent n. 739 saying no and 3 per cent n. 23 unsure). We use

---

6   Alongside the categorisation of sexual identities using quantitative tools, where these tools force/create categorisations and identities rather than simply reflecting them (see Browne 2008, 2011).

trans/non-trans for this categorisation, which places cisgendered people as lacking and in the 'non' category. As we didn't use terms such as 'gender queer' or cisgendered, these are not accurate descriptions of this category. Despite the complexities of defining trans identities, such categories can be useful for attending to key differences between those who identify with and were willing to tick the yes box to this question and those who were not.[7]

Intersectionalities and Multiple Marginalisation(s)

Identity politics have also been critiqued for the ways in which these identities mask differences and do not recognise the intersectional formations of social difference. In particular, sex/gender and sexualities are understood as interdependent with and reformed through other social differences, such as class, race and dis/ability (see for example Fütty 2010, Kuntsman and Miyake 2008, Taylor et al. 2011). Discussions of intersectionality initially were undertaken within feminisms, where black, lesbian and working class women illustrated how the category of woman is not only internally diverse, but created through sexuality, class, race and multiple other social differences (Bryson 1999, Calhoun 1995, The Combahee River Collective 1977, Mills 1997, Mohanty 1992, Sandoval 1991, Wheelan 1995). These discussions have questioned models that simply 'added on' different aspects of identities to woman, without acknowledging how with this 'addition', the category of 'woman' is fundamentally remade (Flax 1990, hooks 1993, Wilton 1995).[8] Identities cannot then be seen as being pieces that are collected in a jigsaw-like way rather intersectionality requires us to see how social differences recreate each other. This happens in a plethora of ways that reconstitute lived experiences, so that these social differences reform each other.[9] A discussion of

---

7   For sexuality, as with most lesbian and gay [*sic*] surveys, gay men form the majority of this sample. Here the breakdown of the sample is 53 per cent gay male, 34 per cent lesbians/gay women (i.e. those who defined as either lesbians or gay and identified as female in the gender question), 6 per cent bisexual and 3 per cent queer (see www.countmeintoo. co.uk for the full breakdown). 'Questioning' and 'other' added up to just under one per cent. Straight/heterosexuals were only included in the overall sample if they indicated that they have had sex/relationships with members of the same sex, or were trans. When this was not the case, they were removed from the sample.

8   It is the 'add-on model' of multiple differences that underpins the 'dual discrimination' clause in the UK's 2010 Equality Act. For Monro (2010) there is evidence of a shift towards an intersectional approach to equalities, but there is an absence of rhetoric of intersectionality. A full discussion of intersectionality and the limits and possibilities of the term 'dual discrimination' is beyond the scope of this chapter (see Taylor et al. 2011).

9   Indeed if we argue that place is active as a social relation there is more work to be done in considering the intersectionalities of place, beyond space as an effect of othering or a manifestation/driver of exclusion, in particular considering how place interacts with and co-constitutes diverse assemblages of identities and practices.

the complexities of intersectionality is beyond the scope of this book (see Taylor et al. 2011). Here, we deploy the term intersectionality to recognise the ways in which social differences recreate each other. However, in this research we were more interested in multiple marginalisations.

We use the concept of multiple marginalisation to explore social differences between LGBT people. In the CMIT research, we used this term to access those who may not feel 'part of' or 'welcome in' the 'gay capital of the UK'. This accounts for experiences of marginalisation because of (heteronormative) sexual and gender identities *as well as* marginalisation from LGBT/gay 'communities' and spaces. In this way, multiple marginalisation draws on intersectional understandings that see identities as assemblages (Puar 2007), which are not neatly delineated into singular categories. In addition, it allows us to discuss what might be termed intersectional differences, for example the classed and gender privilege that has been afforded, and is available, to some lesbians and gay men (see Chapters 4 and 8).

The term 'multiple marginalisation' also enables us to explore hierarchies within categories that are not necessarily 'intersectional' per se. This includes discriminations and differences in experiences between L, G, B and T (in that they are not intersections of multiple identities). Research on hierarchies between these categories offers insights into the differential ways of accessing processes of legitimisation and inclusion. For example, bi people and trans people often existed in a 'middle ground'. They were both accepted in terms of legislation and also marginalised from dominant and ideals of LGBT as seen through only lesbian and gay binaries (see Chapter 5). Thus, we at times use gay and lesbian specifically to highlight the ways in which these subject positions were normalised in ways that trans and bi subjectivities were not.

The CMIT project sought to address the under-engagement of some individuals and groups within LGBT communities and reported on a range of expected and unexpected identities and practices that characterised experiences of marginalisation. Using the term multiple marginalisation afforded us the opportunity to examine practices that do not neatly delineate along (intersectional) identity lines, such as 'not drinking' (see Chapter 4).

In sum, we use LGBT recognising that these identities are contingent and performative. Nonetheless we understand the possibilities of deploying these identities both individually in terms of lesbian, gay, bisexual and trans, and collectively as LGBT. Recognising the differences between those encompassed, or who define themselves, in this category we recognise how intersectionalities structure everyday lives and activisms. However, we more focus more on multiple marginalisations, which are formed in part through and because of intersectional differences.

Bibliography

Adams, J., McCreanor, T. and Braun, V. 2008. Alcohol and Gay Men: Consumption, Promotion and Policy Responses. *Out in Psychology: Lesbian, Gay, Bisexual, Trans and Queer Perspectives*, edited by V. Clarke and E. Peel. West Sussex, England: John Wiley & Sons.

Adler, S. and Brenner, J. 1992. Gender and Space: Lesbians and Gay Men in the City. *International Journal of Urban and Regional Research*, 16, 24–34.

Ahmed, S. 2012. *On Being Included: Racism and Diversity in Institutional Life*. London: Duke.

Aldrich, R. 2004. Homosexuality and the City: An Historical Overview. *Urban Studies*, 41, 1719–1737.

Allen, J. and Cochrane, A. 2010. Assemblages of State Power: Topological Shifts in the Organization of Government and Politics. *Antipode*, 42, 1071–1089.

Allen, J. 2003. *Lost Geographies of Power*. Oxford: Blackwell.

Alsayyad, N. and Roy, A. 2006. Medieval Modernity: On Citizenship and Urbanism in a Global Era. *Space and Polity*, Vol. 10, 1–20.

Amin, A. and Roberts, J. 2008. Knowing in Action: Beyond Communities of Practice. *Research Policy*, 37, 353–369.

Amin, A., Massey, D. and Thrift, N. J. 2002. *Cities: Reimagining the Urban*. Cambridge: Polity Press.

Anderlini-D'Onofrio, S. 2011. Bi Reconnaissance: An Introduction to Bitopia, Selected Proceedings from Birecon. *Journal of Bisexuality*, 11, 146–156.

Anderson, B. 2006. Becoming and Being Hopeful: Towards a Theory of Affect. *Environment and Planning D: Society and Space*, 24, 733–752.

Andrucki, M. J. and Elder, G. 2007. Locating the State in Queer Space: GLBT Non-Profit Organizations in Vermont, USA. *Social and Cultural Geography*, 8, 89–104.

Angelides, S. 2006. Historicising (Bi) Sexuality: A Rejoinder for Gay/Lesbian Studies, Feminism and Queer Theory. *Journal of Homosexuality*, 52, 125–158.

Asthana, S., Richardson, S. and Halliday, J. 2002. Partnership Working In Public Policy Provision: A Framework For Evaluation. *Social Policy and Administration*, 36, 780–795.

Attwood, F. 2010. *Porn.com: Making Sense of Online Pornography*. New York: Peter Lang Publishing.

Attwood, F. 2011. The Paradigm Shift: Pornography Research, Sexualization and Extreme Images. *Sociology Compass*, 5, 13–22.

Barker, M. and Langdridge, D. 2008. II. Bisexuality: Working With a Silenced Sexuality. *Feminism & Psychology*, 18, 389–394.

Barker, M. and Yockney, J. 2004. Including the B-word: Reflections on the Place of Bisexuality within Lesbian and Gay Activism and Psychology. *Lesbian and Gay Psychology Review*, 5, 118–122.

Bassi, C. 2006. Riding the Dialectical Waves of Gay Political Economy: A Story from Birmingham's Commercial Gay Scene. *Antipode: A Radical Journal of Geography*, 38, 213–235.

Bech, H. 1997. *When Men Meet: Homosexuality and Modernity*. Cambridge: Polity Press.

Bell, D. and Binnie, J. 2000. *The Sexual Citizen: Queer Politics and Beyond*. Cambridge: Polity Press.

Bell, D. and Jayne, M. 2004. *City of Quarters: Urban Villages in the Contemporary City*. Aldershot: Ashgate.

Bell, D. and Hollows, J. 2005. *Ordinary Lifestyles: popular media, consumption and taste*. Maidenhead: Open University Press. Bell, D. and Valentine, G. (eds). 1995. *Mapping Desire: Geographies of Sexualities*. London: Routledge.

Bell, D., Binnie, J., Cream, J. and Valentine, G. 1994. All Hyped Up and No Place to Go. *Gender, Place and Culture*, 1, 31–47.

Berlant, L. 2011. *Cruel Optimism*. Durham: Duke.

Binnie, J. and Skeggs, B. 2004. Cosmopolitan Knowledge and the Production and Consumption of Sexualized Space: Manchester's Gay Village. *The Sociological Review*, 52, 39–61.

Binnie, J. 1997. Coming out of Geography: Towards a Queer Epistemology. *Environment and Planning D: Society and Space*, 15, 223–37.

Binnie, J. 2004. *The Globalisation of Sexuality*. London: Sage.

Binnie, J. 2004a. Quartering Sexualities: Gay Villages and Sexual Citizenship. *City of Quarters: Urban Villages in the Contemporary City*, edited by D. F. Bell and M. Jayne. Aldershot: Ashgate.

Binnie, J. 2011. Class, Sexuality, and Space: A Comment. *Sexualities*, 14, 21–26.

Bourdieu, P. 1986. The Forms of Capital. *Handbook of Theory and Research for the Sociology of Education*, edited by J. Richardson. New York: Greenwood.

Brickell, C. 2001. The Transformation of Heterosexism and Its Paradoxes. *Thinking Straight: The Power, the Promise and the Paradox of Heterosexuality*, edited by C. Ingraham. London: Routledge.

Brown, G. 2007. Mutinous Eruptions: autonomous spaces of radical queer activism. *Environment and Planning A*, 39, 2685–2698.

Brown, G. 2008. Urban (Homo)Sexualities: Ordinary Cities and Ordinary Sexualities. *Geography Compass*, 2, 1215–1231.

Brown, G. 2009. Thinking Beyond Homonormativity: Performative Explorations of Diverse Gay Economies. *Environment and Planning A*, 41, 1496–1510.

Brown, G. 2011. Provincial Gay Life: Rethinking Ordinary Homosexualities and Ordinary Cities from the English Midlands. *European Geographies of Sexualities Conference*. Brussels.

Brown, G. 2012. Homonormativity: A Metropolitan Concept that Denigrates "Ordinary" Gay Lives. *Journal of Homosexuality*, 59, 1065–1072.

Brown, M. and Knopp, L. 2003. Queer Cultural Geographies – We're Here! We're Queer! We're Over There, Too! *Handbook of Cultural Geography*, edited by K. Anderson et al. London: Sage.

Brown, M. 2000. *Closet Space: Geographies of Metaphors from the Body to the Globe*. London: Routledge.

Browne, K., Bakshi, L. and Law, A. 2009. Positionalities: It's Not About Them, It's About Us. *The Handbook of Social Geography*, edited by S. P. Smith et al. London: Sage.

Browne, K. and Bakshi, L. 2011. We Are Here to Party? Lesbian, Gay, Bisexual and Trans Leisurescapes beyond Commercial Gay Scenes. *Leisure Studies*, 30, 179–196.

Browne, K. and Bakshi, L. 2012. Insider Activists: The Fraught Possibilities of LGBT Activisms from Within *Geoforum*. [Online] Available at: http://www.sciencedirect.com/science/article/pii/S0016718512002242.

Browne, K. and Davis, P. 2008. *Housing: Count Me In Too Additional Analysis Report*. [Online: Spectrum and the University of Brighton]. Available at: http://www.realadmin.co.uk/microdir/3700/File/CMIT_Housing_Report_April_08.pdf [Accessed 11/01/2012].

Browne, K. and Lim, J. 2008a. *Bi People: Count Me In Too Additional Findings Report*. [Online: Spectrum and the University of Brighton]. Available at: http://www.realadmin.co.uk/microdir/3700/File/CMIT_Bi_Report_Dec08.pdf [Accessed 11/01/2013].

Browne, K. and Lim, J. 2008b *Community Safety: Count Me In Too Additional Analysis Report*. [Online: Spectrum and the University of Brighton]. Available at: http://www.realadmin.co.uk/microdir/3700/File/CMIT_Safety_Report_Final_Feb08.pdf [Accessed 26/02/2011].

Browne, K. and Lim, J. 2008c. *Trans Report: Count Me In Too Additional Analysis Report*. [Online: Spectrum and the University of Brighton]. Available at: http://www.realadmin.co.uk/microdir/3700/File/CMIT_Trans_Report_Dec08.pdf [Accessed 11/01/2013].

Browne, K. and Lim, J. 2010. Trans Lives In The 'Gay Capital of the UK'. *Gender, Place and Culture*, 17, 615–633.

Browne, K. 2006. Challenging Queer Geographies. *Antipode*, 38, 885–893.

Browne, K. 2006a. (Re)Making the Other: Heterosexualising Everyday Space. *Environment and Planning A*, 39, 996–1014.

Browne, K. 2007. *Count Me In Too: Initial Academic Findings Report*. [Online: Spectrum and the University of Brighton]. Available at: http://www.realadmin.co.uk/microdir/3700/File/CMIT_AcademicReport_final_June07.pdf [Accessed 11/01/2013].

Browne, K. 2007a. A Party With Politics?: (Re)Making LGBTQ Pride Spaces in Dublin and Brighton. *Social and Cultural Geography*, 8, 63–87.

Browne, K. 2007b. *Domestic Violence and Abuse Academic Findings Report*. [Online: Spectrum and the University of Brighton]. Available at: http://www.

realadmin.co.uk/microdir/3700/File/CMIT_DV_Report_final_Dec07.pdf [Accessed 11/01/2013].

Browne, K. 2007c. Drag Queens and Drab Dykes: Deploying and Deploring Femininities. *Geographies of Sexualities: Theory, Practices, and Politics*, edited by K. Browne, J. Lim and G. Brown. Aldershot: Ashgate.

Browne, K. 2008. Selling my Queer Soul or Can Queer Research Be Quantitative? *Sociological Research Online*, 13(1). [Online] Available at: http://www. socresonline.org.uk/13/1/11.html.

Browne, K. 2010. Queer Quantification or Queer(y)ing Quantification: Creating Lesbian, Gay, Bisexual or Heterosexual Citizens Through Governmental Social Research. *Queering Methods and Methodologies: Queer Theory and Social Science Methods.* London: Ashgate, pp. 231–249.

Browne, K. 2011. Beyond Rural Idylls: Imperfect Lesbian Utopias at Michigan Womyn's Music Festival. *Journal of Rural Studies*, 27, 13–23.

Browne, K. 2011a. By Partner We Mean … Alternative Geographies Of 'Gay Marriage'. *Sexualities*, 14, 100–122.

Browne, K., Bakshi, L. and Lim, J. 2011. 'It's Something You Just Have to Ignore': Understanding and Addressing Contemporary Lesbian, Gay, Bisexual and Trans Safety Beyond Hate Crime Paradigms. *Journal of Social Policy*, 40, 739.

Browne, K., Bakshi, L. and Lim, J. 2012. 'There's No Point in Doing Research If No One Is Going to Listen': Identifying Lesbian, Gay, Bisexual and Trans Needs and Effecting 'Positive Social Change' for Lesbian, Gay, Bisexual and Trans People in Brighton and Hove. *Social Care, Service Users and User Involvement: Building on Research*, edited by P. Beresford and S. Carr. London: Jessica Kingsley.

Browne, K., Church, A. and Smallbone, K. (2005). Do it with Pride: Lesbian, Gay, Bisexual and Trans Lives and Opinions. University of Brighton. Available at: http://eprints.brighton.ac.uk/5867/1/Do_it_with_Pride_report.pdf.

Browne, K., Cull, M. and Hubbard, P. 2010. The Diverse Vulnerabilities of Lesbian, Gay, Bisexual and Trans Sex Workers in the UK. *New Sociologies of Sex Work*, edited by K. Hardy, S. Kingston and T. Sanders. Ashgate: London.

Browne, K., McGlynn, N. and Lim, J. 2009. *Count Me In Too, LGBT Lives in Brighton and Hove: Drugs and Alcohol Additional Findings Report*. Kath Browne and Spectrum.

Browne, K. and Nash, C. 2010. *Queer Methods and Methodologies*. Aldershot: Ashgate.

Bryant, K. 2008. In Defense of Gay Children? 'Progay' Homophobia and the Production of Homonormativity. *Sexualities*, 11, 455–475.

Bryson, V. 1999. Patriarchy: A Concept Too Useful To Lose. *Contemporary Politics*, 5, 311–324.

Buckland, F. 2002. *Impossible Dance*. Middletown, US: Wesleyan University Press.

Burchill, J. and Raven, D. 2007. *Made in Brighton*. London: Virgin Books.

Butler, J. 1990. *Gender Trouble: Feminism and the Subversion of Identity*. New York: Routledge.

Butler, J. 2004. *Undoing Gender.* London: Routledge.

Cahill, C. 2007. Participatory Data Analysis. *Participatory Action Research Approaches and Methods: Connecting People, Participation and Place*, edited by S. Kindon, R. Pain and M. Kesby. London: Routledge.

Calhoun, C. 1995. The Gender Closet: Lesbian Disappearance under the Sign 'Women'. *Feminist Studies*, 21, 7–34.

Cameron, A. 2007. Progress on Welfare and Exclusion – iii. *Progress in Human Geography*, 31, 519–526.

Campbell, P. and O'Neill, M. 2006. *Sex Work Now.* Cullompton, Devon, UK: Willan Publishing.

Campbell, P. 2002. In Brighton. *London Review of Books*, 24.

Carabine, J. and Monro, S. 2004. Lesbian and Gay Politics and Participation in New Labour's Britain. *Social Politics*, 11, 312–327.

Carmel, E. and Harlock, J. 2008. Instituting the 'Third Sector' As a Governable Terrain: Partnership, Procurement and Peformance in the UK. *Policy and Politics*, 36, 155–171.

Casey, M. 2004. De-dyking Queer Space(S): Heterosexual Female Visibility in Gay and Lesbian Spaces. *Sexualities*, 7, 446–61.

Casey, M. 2007. The Queer Unwanted and their Undesirable Otherness. *Geographies of Sexualities*, edited by K. Browne, J. Lim and G. Brown. Surrey: Ashgate.

Catungal, 2013. Ethno-specific safe houses in the liberal contact zone: Race politics, place-making and the genealogies of the AIDS sector in global-multicultural Toronto. *ACME* forthcoming.

Chatterton, P. A. and Holland, R. 2003. *Urban Nightscapes: Youth Cultures, Pleasure Spaces and Corporate Power*. London: Routledge.

Clarke, N. 2008. From Ethical Consumerism to Political Consumption. *Geography Compass*, 2, 1870–1884.

Clements-Nolle, K., Marx, R. and Katz, M. 2006. Attempted Suicide Among Transgender Persons: The Influence of Gender-Based Discrimination and Victimization. *Journal of Homosexuality*, 51, 53–69.

Clisby, S. 2009. Summer Sex: Youth, Desire and the Carnivalesque at the English Seaside. *Transgressive Sex: Subversion and Control in Erotic Encounters*, edited by H. A. Donnan. Oxford, UK: Bergahn Books.

Cohen, S. 2002. *Folk Devils and Moral Panics: The Creation of the Mods and Rockers.* London: Routledge.

Collins, A. 2004. Sexual Dissidence, Enterprise and Assimilation: Bedfellows in Urban Regeneration. *Urban Studies*, 41, 1789–1806.

Collis, R. 2010. *New Encyclopedia of Brighton.* Brighton: Brighton and Hove Council.

Combahee River Collective, 1979. A Black Feminist Statement. *Capitalist Patriarchy and the Case for Socialist Feminism*, edited by Z. R. Eisenstein. New York: The Monthly Review Press (first published in 1977).

Coogan, K. 2006. Fleshy Specificity: (Re) Considering Transexual Subjects In Lesbian Communities. *Journal of Lesbian Studies*, 10, 17–41.

Cooper, D. and Monro, S. 2003. Governing from the Margins: Queering the State of Local Government? *Contemporary Politics*, 9, 229–255.

Cooper, D. 1994. *Sexing the City; Lesbian and Gay Politics within the Activist State*. London: Rivers Oram.

Cooper, D. 1995. *Power in Struggle: Feminism, Sexuality and the State.* New York: New York University Press.

Cooper, D. 2004. *Challenging Diversity: Rethinking Equality and the Value of Difference*. Cambridge: Cambridge University Press.

Cooper, D. 2006. Active Citizenship and the Governmentality of Local Lesbian and Gay Politics. *Political Geography*, 25, 921–943.

Cooper, D. 2007. Being in Public: The Threat and Promise of Stranger Contact. *Law & Social Inquiry*, 32, 203–232.

Cooper, D. 2009. Caring for Sex and the Power of Attentive Action: Governance, Drama and Conflict in Building a Queer Feminist Bathhouse. *Journal of Women in Culture and Society*, 35, 105–129.

Cooper, D. 2011. Reading the State as a Multi-Identity Formation: The Touch and Feel of Equality Governance. *Feminist Legal Studies*, 19, 3–25.

Cornwall, A. and Jewkes, R. 1995. What is Participatory Research? *Society, Science and Medicine*, 41, 1667–1776.

Craig, G. and Taylor, M. 2002. Dangerous Liaisons: Local Government and the Voluntary and Community Sectors. *Partnerships, New Labour and the Governance of Welfare*, edited by K. Rummery, C. Glendinning and M. Powell. Bristol: Policy Press.

Cull, M., Platzer, H. and Balloch, S. 2006. Out On My Own: Understanding the Experiences and Needs of Homeless Lesbian, Gay, Bisexual and Transgender Youth. Brighton: University of Brighton. [Available at: https://login.live.com/login.srf?wa=wsignin1.0&rpsnv=11&ct=1376386587&rver=6.1.6206.0&wp=MBI&wreply=http:%2F%2Fmail.live.com%2Fdefault.aspx&lc=2057&id=64855&mkt=en-gb&cbcxt=mai&snsc=1] [Accessed 11/01/2013].

D'Emilio, J. 1983. Capitalism and Gay Identity. *Powers of Desire: The Politics of Sexuality*, edited by A. Snitow, C. Stansell and S. Thompson. New York Monthly Review.

Delany, S. 2001. *Times Square Red, Times Square Blue*. New York: New York University Press.

Del Casino, V. 2011. US Social Geography, alive and well? *Social & Cultural Geography* 12: 538-543.

Doan, P. 2007. Queers in the American City: Transgendered Perceptions of Urban Space. *Gender, Place and Culture*, 14, 57–74.

Doan, P. 2009. Safety and Urban Environments: Transgendered Experiences of the City. *Women and Environment*, 78/79.

Doan, P. 2010. The Tyranny of Gendered Spaces: Living Beyond the Gender Dichotomy. *Gender, Place and Culture*, 17, 635–654.

Duggan, L. 2002. The New Homonormativity: The Sexual Politics of Neoliberalism. *Materializing Democracy: Toward a Revitalized Cultural Politics*, edited by R. Castronovo and D. D. Nelson. Durham: Duke University Press.

Duncan, S. and Smith, D. 2006. Individualisation versus the Geography of 'New' Families. *Twenty-First Society*, 1, 167–89.

Elder, G. 2002. Response to 'Queer Patriarchies, Queer Racisms, International'. *Antipode*, 34, 988–91.

Equalities and Human Rights Commission. 2013. Transgender what the law says. Equalities and Human Rights Commission, http://www.equalityhumanrights. com/advice-and-guidance/your-rights/transgender/transgender-what-the-law-says/ Accessed 06/08/13.

Erel, U., Haritaworn, J., Rodríguez, E. N. G. R. and Klesse, C. 2011. On the Depoliticisation of Intersectionality Talk: Conceptualising Multiple Oppressions in Critical Sexuality Studies. *Theorizing Intersectionality and Sexuality (Genders and Sexualities in the Social Sciences)*. Edited by Y. Taylor, S. Hines and M. E. Casey. Basingstoke: Palgrave.

Evans, D. 1993. *Sexual Citizenship: The Material Construction of Sexualities*. London and New York: Routledge.

Faderman, L. 1992. The Return of the Butch/Femme; a Phenomenon of Lesbian Sexuality in the 1980s and 1990s. *Journal of the History of Sexuality*, 2, 578–596.

Fincher, R. and Jacobs, J. M. 1998. *Cities of Difference*. New York: The Guildford Press.

Flax, J. 1990. *Disputed Subjects. Essays on Psychoanalysis, Politics and Philosophy.* New York: Routledge.

Florida, R. L. 2004. *The Rise of the Creative Class and How It's Transforming Work, Leisure, Community and Everyday Life*. New York: Basic Books.

Fuller, C. and Geddes, M. 2008. Urban Governance Under Neoliberalism: New Labour and the Restructuring of State-Space. *Antipode*, 40, 252–282.

Futty, J. T. 2010. Challenges Posed By Transgender – Passing Ambiguities and Interrelations. *Graduate Journal Social Science*, 7, 57–75.

Gammon, M. and Isgro, K. I. 2006. Troubling the Canon: Bisexuality and Queer Theory. *Journal of Homosexuality*, 52, 159–184.

Gatenby, B. and Humphries, M. 2000. Feminist Participatory Action Research: Methodological and Ethical Issues. *Women's Studies International Forum*, 23, 89–105.

Gibson, E. 1978. Understanding the Subjective Meaning of Places. *Humanistic Geography: Prospects and Problems*, edited by D. Ley and M. Samuels. London: Croom Helm.

Giffney, N. 2004. Denormatizing Queer Theory: More than (Simply) Lesbian and Gay Studies. *Feminist Theory*, 5, 73–8.

Gorman-Murray, A. 2007. Reconfiguring Domestic Values: Meanings of Home for Gay Men and Lesbians. *Housing, Theory and Society*, 24, 229–246.

Gorman-Murray, A. 2008. Queering the Family Home: Narratives from Gay, Lesbian and Bisexual Youth Coming Out in Supportive Family Homes in Australia. *Gender, Place and Culture*, 15, 31–44.

Gorman-Murray, A. 2008a. Reconciling Self: Gay Men and Lesbians Using Domestic Materiality for Identity Management. *Social and Cultural Geography*, 9, 283–301.

Gorman-Murray, A. 2009. What's the Meaning of Chillout? Rural/Urban Difference and the Cultural Significance of Australia's Largest Rural GLBTQ Festival. *Rural Society*, 19, 71–86.

Gorman-Murray, A., Waitt, G. and Gibson, C. 2012. Chilling Out in 'Cosmopolitan Country': Urban/Rural Hybridity and the Construction of Daylesford as a 'Lesbian and Gay Rural Idyll', *Journal of Rural Studies*, 28 (1), 69–79.

Gough, P. 2005. Brighton as a Place of Delirious Invention. *The Brighton Book*. Brighton: Myriad City Arts.

Green, R. 2010. Transsexual Legal Rights in the United States and United Kingdom: Employment, Medical Treatment, and Civil Status. *Archives of Sexual Behavior*, 39, 153–160.

Grosz, E. 1998. Bodies-Cities. *Places through the Body*, edited by H. J. Nast and S. Pile. London: Routledge.

Grundy, J. and Smith, M. 2007. Activist Knowledges in Queer Politics. *Economy and Society*, 36, 294–317.

Halberstam, J. 1998. *Female Masculinity*. London: Duke University Press.

Halberstam, J. 2005. *In a Queer Time and Place: Transgender Bodies, Subcultural Lives*. London: New York University Press.

Halberstam, J. 2011. *The Queer Art of Failure*. London: Duke.

Halkitis, P. N. Palamara, J. J. and Mukherjee, P. P. 2007. Poly-Club-Drug Use among Gay and Bisexual Men: A Longitudinal Analysis. *Alcohol Dependence*, 89, 153–160.

Haraway, D. 1991. *Simians, Cyborgs and Women: The Reinvention of Nature*. London: FA.

Hemingway, J. 2006. Sexual Learning and the Seaside: Relocating the Dirty Weekend and Teenage Girls Sexuality. *Sex Education: Sexuality, Society and Learning*, 6, 429–443.

Hemmings, C. 2002. *Bisexual Spaces: A Geography of Sexuality and Gender*. New York and London: Routledge.

Herek, G. M., Gillis, J. R., Cogan, J. C. and Glunt, E. K. 1997. Hate Crime Victimization among Lesbian, Gay, and Bisexual Adults – Prevalence, Psychological Correlates, and Methodological Issues. *Journal of Interpersonal Violence*, 12, 195–215.

Highmore, B. 2002. *Everyday Life and Cultural Theory*. London: Routledge.

Highmore, B. 2010. *Ordinary Lives: Studies in Everyday Life*. London: Routledge.

Hines, S. 2007. *Transforming Gender: Transgender Practices of Identity, Intimacy and Care*. Bristol: Policy Press.

Hines, S. 2010. Queerly Situated? Exploring Negotiations of Trans Queer Subjectivities at Work and within Community Space. *Gender, Place and Culture*, edited by S. Hines. London: Routledge.

Holt, M. and Griffin, C. 2003. Being Gay, Being Straight, and Being Yourself: Local and Global Reflections on Identity, Authenticity and the Lesbian and Gay Scene. *European Journal of Cultural Studies*, 6, 404–425.

hooks, B. 1993. *Sisters of the Yam: Black Women and Self-Recovery*. Boston, MA: South End Press.

Hubbard, P. 2002. Sexing the Self: Geographies of Engagement and Encounter. *Social and Cultural Geography*, 3 (4), 365–381.

Hubbard, P. 2004. Revenge and Injustice in the Neoliberal City: Uncovering Masculinist Agendas. *Antipode*, 665–686.

Hubbard, P. J. 2006. *The City – Key Ideas in Geography*. London: Routledge.

Hubbard, P. J. 2011. *Cities and Sexualities*. London: Routledge.

Hubbard, P. 2011a. Redundant? Resurgent? Relevant? Social Geography in Social & Cultural Geography. *Social & Cultural Geography*, 12 (6), 529–533.

Hughes, H. L. 2003. Marketing Gay Tourism in Manchester: New Market for Urban Tourism or Destruction of 'Gay Space'? *Journal of Vacation Marketing*, 9, 152–163.

Hughes, H. L. 2006. Gay and Lesbian Festivals: Tourism in the Change from Politics to Party. *Festivals, Tourism and Social Change: Remaking Worlds*, edited by D. Picard and M. Robinson. Cleveden: Channel View Publications.

Hughes, H. L. 2006. *Pink Tourism: Holidays of Gay Men and Lesbians*. Oxford: CABI.

Hutta, J. S. 2009. Geographies of Geborgenheit: Beyond Feelings of Safety and the Fear of Crime. *Environment and Planning D: Society and Space*, 27, 251–73.

Jagose, A. 1996. *Queer Theory: An Introduction*. Melbourne: University of Melbourne Press.

Jayne, M., Valentine, G. and Holloway, S. L. 2010. Emotional, Embodied and Affective Geographies of Alcohol, Drinking and Drunkenness. *Transactions of the Institute of British Geographers*, 35, 540–554.

Jayne, M., Valentine, G. and Holloway, S. L. 2011. *Alcohol, Drinking, Drunkenness: (Dis)Orderly Spaces*. Aldershot: Ashgate.

Jenks, C. 2003. *Transgression: Key Ideas*. London and New York: Routledge.

Jessop, B. 1990. *State Theory: Putting the Capitalist State in Its Place*. Cambridge: Polity.

Johnson, K. 2007. Changing Sex, Changing Self. *Men and Masculinities*, 10, 54–70.

Johnson, K. 2007a. Fragmented Identities, Frustrated Politics: Transexuals, Lesbians and Queer. *Twenty-First Century Lesbian Studies*, edited by N. Giffney and K. O'Donnell. New York: Harrington Park Press.

Johnston, L. 1998. Queen(s') Street or Ponsonby Poofters? Embodied HERO Parade Sites. *New Zealand Geographer*, 53, 29–33.

Johnston, L. 2001. (Other) Bodies and Tourism Studies. *Annals of Tourism Research*, 28, 180–201.

Johnston, L. 2005. *Queering Tourism: Paradoxical Performances at Gay Pride Parades*. London: Routledge.

Joloza, T., Evans, S. T. and O'Brian, R. 2010. *Measuring Sexual Identity: An Evaluation Report*. London, UK: Office for National Statistics.

Jones, O. 2011. *Chavs: The Demonization of the Working Class*. London: Verso.

Jorm, A., Korten, A., Rodgers, B., Jacomb, P. and Christensen, H. 2002. Sexual Orientation and Mental Health: Results from a Community Survey of Young and Middle-Aged Adults. *British Journal of Psychiatry*, 180, 423–7.

Kapoor, I. 2005. Participatory Development, Complicity and Desire. *Third World Quarterly*, 26, 1203/1220.

Kates, S. and Belk, R. 2001. The Meanings of Lesbian and Gay Pride Day: Resistance through Consumption and Resistence to Consumption. *Journal of Contemporary Ethnography*, 30, 392–429.

Kennedy, E. L. and Davis, M. 1993. *Boots of Leather, Slippers of Gold: The History of a Lesbian Community*. New York: Routledge.

Kesby, M. 2007. Spatialising Participatory Approaches: The Contribution of Geography to a Mature Debate. *Environment and Planning A*, 39, 2813–2831.

Kindon, S. Pain, R. and Kesby, M. 2008. *Participatory Action Research Approaches and Methods: Connecting People, Participation and Place*. London: Routledge.

Kipke, M. D. et al. 2007. Club Drug Use in Los Angeles Among Young Men Who Have Sex with Men. *Use Misuse*, 42, 1723–1743.

Knopp, L. 1990. Exploiting the Rent-Gap: The Theoretical Significance of Using Illegal Appraisal Schemes to Encourage Gentrification in New Orleans. *Urban Geography*, 11, 48–64.

Knopp, L. 1992. Sexuality and the Spatial Dynamics of Capitalism. *Environment and Planning D: Society and Space*, 10, 651–669.

Knopp, L. 1998. Sexuality and Urban Space: Gay Male Identity Politics in the United States, the United Kingdom, and Australia. *Cities of Difference*, edited by J. Jacobs and R. Fincher. London: The Guildford Press.

Knopp, L. 2004. Ontologies of Place, Placelessness, and Movement: Queer Quests for Identity and Their Impacts on Contemporary Geographic Thought. *Gender, Place and Culture*, 11, 121–134.

Koch, R. and Latham, A. 2012. Rethinking Urban Public Space: Accounts From a Junction in West London. *Transactions of the Institute of British Geographers*, 37 (4), 515–529.

Kramer, J. L. 1995. Bachelor Farmers and Spinsters: Gay and Lesbian Identities and Communities in Rural North Dakota. *Mapping Desire*, edited by D. Bell and G. Valentine. New York: Routledge.

Kuntsman, A. and Miyake, E. 2008. *Out of Place: Interrogating Silences in Queerness/Raciality.* York, UK: Raw Nerve Books.

Laing, S. and Maddison, E. 2007. The CUPP Model In Context. *Community–University Partnerships in Practice*, edited by A. Hart, D. Wolff and E. Maddison. Leicester: Niace.

Lauria, M. and Knopp, L. 1985. Towards an Analysis of the Role of Gay Communities in the Urban Renaissance. *Urban Geography*, 6, 152–69.

Lees, L. 2004. *The Emancipatory City: Paradoxes and Possibilities.* London: Sage.

Levine, M. P. 1979. Gay Ghetto. *Gay Men: The Sociology of Male Homosexuality*, edited by M. Levine. New York: Harper & Row.

Lim, J. and Browne, K. 2009. Senses of Gender. *Sociological Research Online*, 14.

Lim, J. 2007. Queer Critique and the Politics of Affect. *Geographies of Sexuality: Theory, Practices and Politics*, edited by K. Browne, J. Lim and G. Brown. Aldershot: Ashgate.

Lloyd, M. 2005. Butler, Antigone and State. *Contemporary Political Theory*, 4, 451–468. London: Pluto Press.

Markwell, K. 2002. Mardi Gras Tourism and the Construction of Sydney as an International Gay and Lesbian City. *GLQ: A Journal of Lesbian and Gay Studies*, 8, 81–99.

Martin, A. K. 1997. The Practice of Identity and an Irish Sense of Place. *Gender, Place and Culture – A Journal of Feminist Geography*, 4, 89–114.

Mason, G. 2007. Hate Crime as Amoral Category: Lessons from the Snowtown Case, Australian and New Zealand. *Journal of Criminology*, 40, 249–71.

Massey, D. 2005. *For Space.* London: Sage.

May, J. 2003. Local Connection Criteria and Single Homeless People's Geographical Mobility: Evidence from Brighton and Hove. *Housing Studies*, 18, 29–46.

McCabe, S. E., Bostwick, W. B., Hughes, T. L., West, B. T. and Boyd, C. J. 2010. The Relationship between Discrimination and Substance Use Disorders among Lesbian, Gay, and Bisexual Adults in the United States. *American Journal of Public Health*, 100, 1946–1952.

McDermott, E. 2011. The World Some Have Won: Sexuality, Class and Inequality. *Sexualities*, 14, 63–78.

McDermott, E. 2011a. Multiplex Methodologies: Researching Young People's Well-Being at the Intersections of Class, Sexuality, Gender and Age. *Theorising Intersectionality and Sexuality*, edited by S. Hines, M. Casey and Y. Taylor. Basingstoke: Palgrave Macmillan.

McDowell, L. 1999. *A Feminist Glossary of Human Geography.* London: Arnold.

McFarlane, C. 2011. *Learning the City: Knowledge and Translocal Assemblage.* Oxford: Wiley-Blackwell.

McGhee, D. 2003. Joined-Up Government, 'Community Safety' and Lesbian, Gay, Bisexual and Transgender 'Active Citizens'. *Critical Social Policy*, August, 23, 345–374.

McGhee, D. 2004. Beyond Toleration: Privacy, Citizenship and Sexual Minorities in England and Wales. *The British Journal of Sociology*, 55, 357–375.

McLean, K. 2007. Hiding in the Closet? *Journal of Sociology*, 43, 151–166.

McNulty, A., Richardson, D. and Monro, S. 2010. *Lesbian, Gay, Bisexual and Trans (LGBT) Equalities and Local Governance: Research Report for Practitioners and Policy Makers*. Newcastle: University of Newcastle.

Memon, A. and Walker, A. 2008. *Brighton and Hove Suicide Prevention Strategy 2008–2011*. NHS Brighton and Hove.

Mills, S. 1998. Post-colonial feminist theory. *Contemporary Feminist Theories*, edited by Jackson, S. and Jones, J. Edinburgh: Edinburgh University Press.

Mohanty, C. T. 1992. Feminist Encounters: Locating The Politics of Experience. *Destabilizing Theory: Contemporary Feminist Debates*, edited by M. Barrett and A. Phillips. Cambridge: Polity Press.

Monro, S. 2003. Transgender Politics in the UK. *Critical Social Policy*, 23, 433–452.

Monro, S. 2005. *Gender Politics: Activism, Citizenship and Sexual Diversity*. London: Pluto Press.

Monro, S. 2006. Evaluating Local Government Equalities Work: The Case of Sexualities Initiatives in the UK. *Local Government Studies*, 32, 19–39.

Monro, S. 2007. New Institutionalism and Sexuality at Work in Local Government. *Gender, Work and Organization*, 14, 1–19.

Monro, S. 2010. Sexuality, Space and Intersectionality: The Case of Lesbian, Gay and Bisexual Equalities Initiatives in UK Local Government. *Sociology*, 44, 996–1010.

Moran, L. J. 2001. Affairs of the Heart: Critical Reflections on Hate Crime. *Law and Critique*, 12, 1–15.

Moran, L. J., Skeggs, B., Tyler, P. and Corteen, K. 2004. *Sexuality and the Politics of Violence and Safety*. London: Routledge.

Moran, L. J. and Sharpe, A. N. 2004. Violence, Identity and Policing: The Case of Violence Against Transgender People. *Criminal Justice*, 4, 395–417.

Morgan, J. 2002. US Hate Crime Legislation: A Legal Model to Avoid in Australia. *Journal of Sociology*, 38, 25–48.

Mowlabocus, S. 2010. *Gaydar Culture: Gay Men, Technology and Embodiment in the Digital Age*. Abingdon, UK: Ashgate.

mrs kinpaisby. 2008. Taking Stock of Participatory Geographies: Envisioning the Communiversity. *Transactions of the British Geographers*, 33, 292–299.

Munt, S. 1995. The Lesbian Flaneur. *Mapping Desire: Geographies of Sexualities*, edited by D. Bell and G. Valentine. London: Routledge.

Munt, S. R. 2007. *Queer Attachments: The Cultural Politics of Shame*. London: Ashgate.

Myslik, W. 1996. Renegotiating the Social/Sexual Identities of Places. *Bodyspace: Destabilizing Geographies of Gender and Sexuality*, edited by N. Duncan. London: Routledge.

Namaste, V. 2000. *Invisible Lives: The Erasure of Transexual and Transgendered People*. Chicago and London: University of Chicago Press.

Nash, C. 2006. Toronto's Gay Village (1969–1982): Plotting the Politics of Gay Identity. *The Canadian Geographer*, 50, 1–16.

Nash, C. 2010. Trans Geographies, Embodiment and Experience. *Gender, Place and Culture*, 17, 579–595.

Nash, C. J. 2011. Trans Experiences in Lesbian and Queer Space. *Canadian Geographer/Le Géographe Canadien*, 55, 192–207.

Nast, H. 2002. Queer Patriarchies, Queer Racisms, International. *Antipode*, 34, 874–909.

Newman, J. and Clarke, J. 2009. *Publics, Politics and Power: Remaking the Public in Public Services*. London: Sage.

Noble, G. 2012. Spaces of Privilege (PhD Thesis). Brighton: University of Brighton.

Noble, J. B. 2006. *Sons of the Movement: FTMs Risking Incoherence in a Post-Queer Political Landscape*. Toronto: Women's Press.

O'Brien, J. 2008. Afterword: Complicating Homophobia. *Sexualities*, 11, 496–512.

Ong, A. 2006. *Neoliberalism as Exception: Mutations in Citizenship and Sovereignty*. United States: Duke University Press.

Oswin, N. 2005. Towards Radical Geographies of Complicit Queer Futures. *ACME: An International E-Journal for Critical Geographers*, 3, 79–86.

Oswin, N. 2008. Critical Geographies and the Uses of Sexuality: Deconstructing Queer Space. *Progress in Human Geography*, 32, 89–103.

Oxley, E. and Lucius, C. 2000. Looking Both Ways: Bisexuality and Therapy. *Pink Therapy: Issues in Therapy with Lesbian, Gay and Bisaexual Clients*, edited by C. Neal and D. Davies. Buckingham: Oxford University Press.

Pain, R. and Francis, P. 2003. Reflections on Participatory Research. *Area*, 35, 46–54.

Pain, R. 2004. Social Geography: Participatory Research. *Progress in Human Geography*, 28, 652–63.

Painter, J. 2006. Prosaic Geographies Of Stateness. *Political Geography*, 25, 752–774.

Peel, E. and Harding, R. 2008. Editorial Introduction: Recognizing and Celebrating Same-Sex Relationships: Beyond the Normative Debate. *Sexualities*, 11, 659–666.

Perry, B. 2001. *In the Name of Hate: Understanding Hate Crimes*. New York: Routledge.

Petridis, A. 2010. Is Brighton Britain's Hippest City? *The Guardian*, Wednesday 19 May 2010.

Phillips, R., Watt, D. and Shittleton, D. 2000. *Decentering Sexualities: Politics and Representations Beyond the Metropolis*. New York and London: Routledge.

Podmore, 2013. Lesbians as Village 'Queers': The Transformation of Montréal's Lesbian Nightlife in the 1990s. *ACME* forthcoming.

Podmore, J. 2001. Lesbians in the crowd: Gender, sexuality, and visibility along Montréal's Boulevard St.-Laurent. *Gender, Place, and Culture*, 191–217.

Powell, M. and Dowling, B. 2006. New Labour's Partnerships: Comparing Conceptual Models with Existing Forms. *Social Policy and Society*, 5, 305–314.

Preston-Whyte, R. The Beach as a Liminal Space. *A Companion to Tourism*, edited by A. A. Lew, C. Hall and A. M. Williams. Oxford: Blackwell.

Pritchard, A., Morgan, N. J. And Sedgley, D. 2002. In Search of Lesbian Space? The Experience of Manchester's Gay Village. *Leisure Studies*, 21, 105–123.

Pritchard, A., Morgan, N. J., Sedgely, D. and Jenkins, A. 1998. Reaching Out To the Gay Tourist: Opportunities and Threats in an Emerging Market Segment. *Tourism Management*, 19, 273–282.

Probyn, E. 2005. *Blush: Faces of Shame*. Minneapolis: University of Minnesota Press.

Prosser, J. 1998. *Second Skins: The Body Narratives of Transsexuality*. New York: Columbia University.

Puar, J. K. 2002. A Transnational Feminist Critique of Queer Tourism. *Antipode*, 34, 935–46.

Puar, J. K. 2007. *Terrorist Assemblages: Homonationalism in Queer Times*. United States: Duke University Press.

Queer Mutiny, 2010a. Queer Mutiny Manifesto [Online]. Available at: http://s3.amazonaws.com/jef.mindtouch.com/104266/15/0?Awsaccesskeyid=1TDEJCXAPFCDHW56MSG2&Signature=0czqn8jmznyma9taw1qbsi1wyvu%3d&Expires=1290847180 [Accessed 15th March 2010].

Queer Mutiny. 2010b. *Queer Mutiny Website* [Online]. Available at: http://queermutinybrighton.wik.is/ [Accessed 15/03/2011].

Quilley, S. 1995. Manchester's 'Village in the City': The Gay Vernacular in a Post-Industrial Landscape of Power. *Transgressions: A Journal of Urban Exploration*, 1, 36–50.

Rayside, D. 1998. *On the Fringe – Gays and Lesbians in Politics*. New York: Cornell University.

Relph, E. 1976. *Place and Placelessness*. London: Pion.

Richardson, D. and May, H. 1999. Deserving Victims?: Sexual Status and the Social Construction of Violence. *Sociological Review*, 47, 308–331.

Richardson, D. and Monro, S. 2010. Intersectionality and Sexuality: The Case of Sexuality and Transgender Equalities Work in UK Local Government. *Theorising Intersectionality and Sexuality*, edited by S. Hines, M. Casey and Y. Taylor. Basingstoke: Palgrave Macmillan.

Richardson, D. and Monro, S. 2012. *Sexuality, Equality and Diversity*. Basingstoke: Palgrave.

Richardson, D. 2004. Locating Sexualities: From Here To Normality. *Sexualities*, 7, 391–411.

Richardson, D. 2005. Desiring Sameness? The Rise of a Neoliberal Politics of Normalisation. *Antipode*, 37 (3), 515–535.

Robinson, J. 2002. Global and World Cities: A View from Off the Map. *International Journal of Urban and Regional Research*, 26, 531–554.

Robinson, J. 2005. Urban Geography: World Cities, Or a World of Cities. *Progress in Human Geography*, 29, 757–765.

Robinson, J. 2006. *Ordinary Cities: Between Modernity and Development*. London: Routledge.

Roen, K. 2001. 'Either/Or' and 'Both/Neither': Discursive Tensions in Transgender Politics. *Journal of Woman in Culture and Society*, 27.

Rose, G. 1993. *Feminism and Geography: The Limits of Geographical Knowledge*. Minneapolis: University of Minnesota Press.

Rose, G. 1995. Distance, Surface, Elsewhere: A Feminist Critique of the Space of Phallocentric Self/Knowledge. *Environment and Planning D: Society and Space*, 13, 761–781.

Rothenberg, T. 1995. 'And She Told Two Friends': Lesbians Creating Urban Social Space. *Mapping Desire: Geographies of Sexuality*, edited by D. Bell and G. Valentine. London: Routledge.

Rummery, K. 2002. Towards a Theory of Welfare Partnerships. *Partnerships, New Labour and the Governance of Welfare*, edited by C. Glendinning, M. Powell and K. Rummery. Bristol: The Policy Press.

Rushbrook, D. 2002. Cities, Queer Space and the Cosmopolitan Tourist. *GLQ: A Journal of Lesbian and Gay Studies*, 183–206.

Ruting, B. 2008. Economic Transformations of Gay Urban Spaces: Revisiting Collins' Evolutionary Gay District Model. *Australian Geographer*, 39, 259–269.

Ryan, S. 2010. Brighton and Hove Named Drugs Death Capital. *The Argus*, 25/08/10.

Said, E. 1978. *Orientalism*. New York: Vintage.

Sandoval, C. 1991. U.S. Third World Feminism: The Theory and Method of Oppositional Consciousness in the Postmodern World. *Genders*, 10, 1–24.

Scott, P. 1998. *Zorro Report: An Assessment of the HIV Prevention Needs of Gay Men in Brighton and the Adequacy of Local HIV Prevention Services in Meeting Those Needs*.

Sears, A. 2005. Queer Anti-Capitalism: What's Left of Lesbian and Gay Liberation? *Science and Society*, 69, 95–112.

Sedgwick, E. K. 2003. *Touching Feeling: Affect, Pedagogy, Performativity*. Durham, NC: Duke University Press.

Seidman, S. 2002. *Beyond the Closet: The Transformation of Gay and Lesbian Life*. London: Routledge.

Shields, R. 1990. The 'System of Pleasure': Liminality and the Carnivalesque at Brighton. *Theory, Culture and Society*, 7, 39–72.

Shields, R. 1991. *Places on the Margin: Alternative Geographies of Modernity.* London: Routledge.

Skeggs, B. 1999. Matter out of Place: Visibility and Sexualities in Leisure Spaces. *Leisure Studies*, 3, 213–232.

Skeggs, B. and Binnie, J. 2004. Cosmopolitan Knowledge and the Production and Consumption of Sexualized Space: Manchester's Gay Village. *The Sociological Review*, 52, 39–61.

Soja, E. W. 1996. *Third Space: Journeys to Los Angeles and Other Real or Imagined Places.* London, UK and Malden, USA: Blackwell.

Soja, E. W. 2010. *Seeking Spatial Justice.* Minneapolis: University of Minnesota Press.

Spade, D. *Normal Life: Administrative Violence, Critical Trans Politics, and the Limits of Law.* New York: South End Press.

Spectrum, 2008 Statement of purpose. www.spectrum-lgbt.org/downloads/ Spectrum_Statement_of_Purpose.doc, no longer available online.

Stenson, K. 2008. Governing the Local: Sovereignty, Social Governance and Community Safety. *Social Work and Society*, 6, 1–14.

Stoecker, R. 1999. Making Connections: Community Organizing, Empowerment Planning, And Participatory Research in Participatory Evaluation. *Sociological Practice*, 1, 209–232.

Stonewall. 2011. *Stonewall Top Employers 2011: The Workplace Equality Index* [Online]. Available at: http://www.stonewall.org.uk/documents/wei_2011_ final_booklet.pdf [Accessed:23/11/11].

Stonewall (2013) http://www.stonewall.org.uk/# [Accessed 16 August 2013].

Stryker, S. 2006. (De)Subjugated Knowledges: An Introduction to Transgender Studies. *The Transgender Studies Reader*, edited by S. Stryker and S. Whittle. London: Routledge.

Stychin, C. F. 2003. *Governing Sexuality.* Oxford: Hart Publishing.

Stychin, C. F. 2006. 'Las Vegas is not where we are': Queer readings of the Civil Partnership Act. *Political Geography*, 25, 899–920.

Sullivan, A. 1997. *Same Sex Marriage, Pro and Con: A Reader.* New York: Vintage.

Szymanski, D. A. 2005. Heterosexism and Sexism as Correlates of Psychological Distress in Lesbians. *Journal of Counselling and Development*, 83, 355–360.

Taylor, G. 1999. Empowerment, Identity and Participatory Research: Using Social Action Research to Challenge Isolation for Deaf and Hard Of Hearing People from Minority Ethnic Groups. *Disability and Society*, 14, 369–384.

Taylor, Y. and Addison, M. 2011. Placing Research: 'City Publics' and the 'Public Sociologist'. *Sociological Research Online*, 16, 6.

Taylor, Y. 2007. *Working-Class Lesbian Life: Classed Outsiders.* Basingstoke: Palgrave Macmillan.

Taylor, Y. 2007a. Brushed Behind the Bike Shed: Working-Class Lesbians' Experiences of School. *British Journal of Sociology of Education*, 28, 349–362.

Taylor, Y. 2007b. 'If Your Face Doesn't Fit …': The Misrecognition of Working-Class Lesbians in Scene Space. *Leisure Studies*, 26, 161–178.

Taylor, Y. 2009. *Lesbian and Gay Parenting: Securing Social and Educational Capital*. Basingstoke: Palgrave Macmillan.

Taylor, Y. Hines, S. and Casey, M. 2010. *Theorizing Intersectionality and Sexuality: Genders and Sexualities in the Social Sciences*. Basingstoke: Palgrave.

Tett, L. 2005. Partnerships, Community Groups and Social Inclusion. *Studies in Continuing Education*, 27, 1–14.

Thomas, S. 2011. London to Brighton: Round the Bend and Over the Edge. *Literary London: Interdisciplinary Studies in the Representation of London* [Online]. Available at: http://www.literaryLondon.org/London-journal/thomas1.html [Accessed 12/12/11].

Thompson. 2010. House Prices in Brighton and Hove Rise at Record Levels. *The Argus*, 03/06/2010.

Tomlinson, A. 2005. *Sport and Leisure Cultures*. Minnesota: University of Minnesota Press.

Tucker, A. 2009. *Queer Visibilities: Space, Identity and Interaction in Cape Town*. Chichester: Blackwell.

Valentine, D. 2007. *Imagining Transgender: An Ethnography of a Category*. London: Duke University Press.

Valentine, G. and Skelton, T. 2003. Finding Oneself, Losing Oneself: The Lesbian and Gay 'Scene' as a Paradoxical Space. *International Journal of Urban and Regional Research*, 27, 849–66.

Valentine, G. 1989. The Geography of Women's Fear. *Area*, 21, 385–390.

Valentine, G. 1993. (Hetero)Sexing Space: Lesbians Perceptions and Experiences of Everyday Places. *Environment and Planning D: Space and Society*, 11, 395–413.

Valentine, G. 1993. Negotiating and Managing Multiple Sexual Identities: Lesbian Time Management Strategies. *Transactions of the Institute of British Geographers*, 18, 237–48.

Valentine, G. 1996. (Re)Negotiating the 'Heterosexual Street': Lesbian Productions of Space. *Body Space: Destabilising Geographies of Gender and Sexualities*, edited by N. Duncan. London: Routledge.

Vidal-Ortiz, S. 2008. 'The Puerto Rican Way Is More Tolerant': Constructions and Uses of 'Homophobia' Among Santeria Practitioners Across Ethno-Racial and National Identification. *Sexualities*, 11, 476–495.

Visser, G. 2003. Gay Men, Leisure Space and South African Cities: The Case of Cape Town. *Geoforum*, 34, 123–137.

Visser, G. 2008. The Homonormalisation of White Heterosexual Leisure Spaces in Bloemfontein, South Africa. *Geoforum*, 39, 1347–1361.

Visser, G. 2013. Challenging the gay ghetto in South Africa: Time to move on? *Geoforum*, forthcoming Available at: http://dx.doi.org/10.1016/j.geoforum.2012.12.013.

Voss, G., Gupta, C., and Browne, K. forthcoming. Balancing on the Fence? Between Queer and Bisexual Theory at Brighton Bifest. *Journal of Homosexuality*.

Waitt, G. and Gorman-Murray, A. 2008. Camp in the Country: Renegotiating Sexuality and Gender through a Rural Lesbian and Gay Festival. *Journal of Tourism and Cultural Change*, 6, 185–200.

Waitt, G. and Markwell, K. 2006. *Gay Tourism: Culture and Context.* Birmington, US: Haworth.

Ward, J. 2003. Producing 'Pride' in West Hollywood: A Queer Cultural Capital for Queers with Cultural Capital. *Sexualities*, 6, 65–94.

Ward, J. 2008. Dude-Sex: White Masculinities and 'Authentic' Heterosexuality among Dudes Who Have Sex with Dudes. *Sexualities*, 11, 414–434.

Ward, K. 2007. Geography and Public Policy: Activist, Participatory, and Policy Geographies. *Progress in Human Geography*, 31, 695–705.

Ward, Z. 2008. *Respectably Queer – Diversity Culture and LGBT Activist Organisations.* Nashville, US: Vanderbilt University Press.

Warner, M. 1993. *Fear of a Queer Planet; Queer Politics and Social Theory.* Minnesota: University of Minnesota Press.

Warner, M. 1999. *The Trouble with Normal: Sex, Politics, and the Ethics of Queer Life.* New York: The Free Press.

Warner, M. 2002. *Publics and Counterpublics.* New York: Zone Books.

Warrington, M. 2001. I Must Get Out: The Geographies of Domestic Violence. *Transactions of the Institute of British Geographers*, 26, 365–382.

Watson, S. and Goodall, A. 2010. *Why It's So Good to Sing in an LGBT Choir.* Paper presented at the Transforming Lesbian, Gay, Bisexual and Trans Lives: Research and Activism for Sexual and Gender Liberation Conference. Brighton.

Webb, D. and Wright, D. 2001. 'Count Me In' findings from: *The Lesbian, Gay, Bisexual and Transgender Community Needs Assessment.* Southampton: University of Southampton.

Webb, D. 2005. Bakhtin at the Seaside. *Theory, Culture and Society*, 22, 121–138.

Weber, G. N. 2008. Using to Numb the Pain: Substance Use and Abuse among Lesbian, Gay, and Bisexual Individuals. *Journal of Mental Health Counseling*, 30, 31–48.

Weeks, J. 1989. *Sex, Politics and Society: The Regulation of Sexuality Since 1800.* London: Longman.

Weeks, J. 2007. *The World We Have Won: The Remaking of Erotic and Intimate Life.* London: Routledge.

Weeks, J. 2008. Regulation, Resistance, Recognition. *Sexualities*, 11, 787–792.

West, P. 2004. *Report into the Medical and Related Needs of Transgender People in Brighton and Hove; The Case for Local Integrated Service.* Brighton: Spectrum and Brighton and Hove City NHS Primary Care Trust.

Weston, K. 1995. Get Thee to a Big City: Sexual Imaginary and the Great Gay Migration. *GLQ: A Journal of Lesbian and Gay Studies*, 2, 253–277.

Wheelan, I. 1995. *Modern Feminist Thought: From Second Wave to 'Postfeminism'*. Edinburgh: Edinburgh University Press.

Whittle, S. and Lewis, T. 2007. Bereavement: A Guide for Transexual and Transgender People and their Loved Ones. Department of Health/NHS.

Whittle, S. and Lewis, T. 2007. Sex Changes? Paradigm Shifts in 'Sex' and 'Gender' Following the Gender Recognition Act? *Sociological Research Online*, 12.

Whittle, S., Turner, L. and Al-Alami, M. 2007. *Engendered Penalties: Transgender and Transsexual People's Experiences of Inequality and Discrimination*. [Online: The Equalities Review 2007]. Available at: http://www.pfc.org.uk/pdf/engenderedpenalties.pdf [Accessed: 11/01/2013].

Wilkins, R. A. 1997. *First National Survey on Transviolence*. New York: GenderPAC. Wilkinson, E. 2009. The Emotions Least Relevant To Politics: Queering Autonomous Activism. *Emotions, Space and Society*, 2, 36–43.

Wilkinson, E. 2011. 'Extreme pornography' and the contested spaces of virtual citizenship, *Social and Cultural Geography*, 12, 493–508.

Williams, R. 1958/2011. Culture is Ordinary. *Cultural Theory: An Anthology*, edited by I. Szeman and T. Kaposy. Chichester: Wiley-Blackwell.

Wilton, T. 1995. *Lesbian Studies: Setting an Agenda.* London: Routledge.

Yarwood, R. 2007. The Geographies of Policing. *Progress in Human Geography*, 31, 447–465.

Index